INSTRUCTOR'S R
to acc

LINEAR ALGEBRA
IDEAS AND APPLICATIONS

RICHARD C. PENNEY
Purdue University

On Line Mathematica Translations
by
WILLIAM EMERSON
Metropolitan State College

On Line Maple Translations
by
ROBERT LOPEZ
Rose-Hulman Institute of Technology

John Wiley & Sons, Inc.
New York • Chichester • Weinheim • Brisbane • Singapore • Toronto

COVER PHOTO   ©Chris Thomaidis/Tony Stone Images/New York, Inc.

Copyright © 1998 by John Wiley & Sons, Inc.

Excerpts from this work may be reproduced by instructors for distribution on a not-for-profit basis for testing or instructional purposes only to students enrolled in courses for which the textbook has been adopted. *Any other reproduction or translation of this work beyond that permitted by Sections 107 or 108 of the 1976 United States Copyright Act without the permission of the copyright owner is unlawful. Requests for permission or further information should be addressed to the Permissions Department, John Wiley & Sons, Inc., 605 Third Avenue, New York, NY 10158-0012.*

ISBN 0-471-24599-2

Printed in the United States of America

10 9 8 7 6 5 4 3 2 1

Printed and bound by Malloy Lithographing, Inc.

Contents

	Errata	3
1	**Systems Of Linear Equations**	**5**
	1.1 Vectors	5
	1.2 The Vector Space of $m \times n$ Matrices	8
	1.3 Systems of Linear Equations	13
	1.4 Gaussian Elimination	14
	1.4.1 Network Flow	17
	1.5 Column Space and Nullspace	18
	1.5.1 Predator-Prey Problems	23
2	**Dimension**	**25**
	2.1 The Test for Linear Independence	25
	2.2 Dimension	29
	2.2.1 Differential Equations	31
	2.3 Applications to Systems	33
3	**Transformations**	**37**
	3.1 Matrix Transformations	37
	3.2 Matrix Multiplication	40
	3.3 Image of a Transformation	43
	3.4 Inverses	46
	3.5 The **LU** Factorization	48
4	**Orthogonality**	**51**
	4.1 Coordinates	51
	4.2 Projections: The Gram-Schmidt Process	52
	4.3 Fourier Series: Scalar Product Spaces	53
	4.4 Orthogonal Matrices	55
	4.5 Least Squares	56

5 Determinants — 59
- 5.1 Determinants — 59
- 5.2 Reduction and Determinants — 60
- 5.3 A Formula for Inverses — 60

6 Eigenvectors — 63
- 6.1 Eigenvectors — 63
  - 6.1.1 Markov Processes — 64
- 6.2 Diagonalization — 66
- 6.3 Complex Eigenvectors — 68
- 6.4 Matrix of a Linear Transformation — 69
- 6.5 Orthogonal Diagonalization — 71

A Mathematica On Line

B Maple On Line

INSTRUCTOR'S RESOURCE MANUAL
to accompany

LINEAR ALGEBRA
IDEAS AND APPLICATIONS

Errata

Section 1.4.1

Exercise 1b is not workable as stated. The traffic flow is never heaviest on North St.

Section 3.1

In Exercise 23, $\int_{-1}^{1} x^4 \, dx = 0$.

Section 4.1

In Exercise 10b, Section 4.1 the given polynomials are independent.

Section 4.5

The last sentence in Exercise 5 should read "Finally, use the formula $P = ae^{bt}$ to predict the population in the year 2010."

In the On Line section, the formula in the remark should have parentheses around the T's:

```
r=c* (1+a*sin(T)+b*cos(T)).^ (-1)
```

Section 5.1

In Exercise 5, $\alpha = \gamma - \delta$, not $\gamma + \delta$.

Section 5.2

The hint for Exercise 11 in the answer section should refer the student to the argument on p. 286, not p. 274.

Section 6.1.1 The answer in the back of the text for Exercise 3c is wrong. It takes 5 products.

The hint for Exercise 19 should say "Assume that $PX = -X$." Actually, the proof suggested here is unnecessarily complicated. A better hint would be "Apply the result of Exercise 17 to $Q = P^2$." See the solution to this Exercise.

Chapter 1

Systems Of Linear Equations

1.1 Vectors

(1) (a) $A - B = (1, 4) \neq C - D = (-2, -8)$, hence not equivalent (b) $A - B = (1, -9) = C - D$, hence equivalent (c) $A - B = (-3, -3, -3) \neq C - D = (-1, -1, -1)$, hence not equivalent (d) $A - B = (0, -1, 1) \neq C - D = (-1, -2, -1)$, hence not equivalent.

(2) (a) $A + B = [2, 6]$, $|A - B| = \sqrt{(3+1)^2 + (4-2)^2} = 2\sqrt{5}$, $|A| = \sqrt{3^2 + 4^2} = 5$, $|B| = \sqrt{(-1)^2 + 2^2} = \sqrt{5}$, $A \cdot B = -3 + 8 = 5$

(b) $A + B = [2, 6, 0]$, $|A - B| = \sqrt{(3+1)^2 + (4-2)^2 + 0^2} = 2\sqrt{5}$, $|A| = \sqrt{3^2 + 4^2 + 0^2} = 5$, $|B| = \sqrt{(-1)^2 + 2^2 + 0^2} = \sqrt{5}$, $A \cdot B = -3 + 8 + 0 = 5$

(c) $A + B = [2, 0, 6]$, $|A - B| = \sqrt{(3+1)^2 + 0^2 + (4-2)^2} = 2\sqrt{5}$, $|A| = \sqrt{3^2 + 0^2 + 4^2} = 5$, $|B| = \sqrt{(-1)^2 + 0^2 + 2^2} = \sqrt{5}$, $A \cdot B = -3 + 0 + 8 = 5$

(d) $A + B = (-7, 9)$, $|A - B| = \sqrt{(1)^2 + (-5)^2} = \sqrt{26}$, $|A| = \sqrt{(-3)^2 + 2^2} = \sqrt{13}$, $|B| = \sqrt{(-4)^2 + 7^2} = \sqrt{65}$, $A \cdot B = 12 + 14 = 26$

(e) $A + B = (9, 12)$, $|A - B| = \sqrt{(-3)^2 + (-4)^2} = 5$, $|A| = \sqrt{3^2 + 4^2} = 5$, $|B| = \sqrt{6^2 + 8^2} = 10$, $A \cdot B = 18 + 32 = 50$

(f) $A + B = (2, 2, 1)$, $|A - B| = \sqrt{0^2 + 0^2 + (-1)^2} = 1$, $|A| = \sqrt{1^2 + 1^2 + 0^2} = \sqrt{2}$, $|B| = \sqrt{0^2 + 1^2 + 1^2} = \sqrt{2}$, $A \cdot B = 0 + 1 + 0 = 1$

(g) $A + B = (1, 3, 2)$, $|A - B| = \sqrt{1^2 + (-1)^2 + (-2)^2} = \sqrt{6}$, $|A| = \sqrt{1^2 + 1^2 + 0^2} = \sqrt{2}$, $|B| = \sqrt{0^2 + 2^2 + 2^2} = 2\sqrt{2}$, $A \cdot B = 0 + 2 + 0 = 2$.

(3) (a) Not perpendicular: $A \cdot B = 14 - 3 = 11 \neq 0$ Perpendicular: (b) $A \cdot B = 3 - 2 - 1 = 0$ (c) Perpendicular: $A \cdot B = -ab + ab = 0$, (d) Perpendicular: $A \cdot B = 6 - 4 - 2 = 0$

(4) From formula (6), $B \cdot A = |B|\sqrt{2}\cos\theta \leq |B|$ if and only if $\cos\theta \leq \frac{\sqrt{2}}{2}$. This describes the *complement* of the first quadrant in R^2.

(5) From formula (6), $B \cdot A = |B|\sqrt{3}\cos\theta \geq \frac{3}{2}|B|$ if and only if $\cos\theta \geq \frac{\sqrt{3}}{2}$. This describes the cone consisting of all vectors which make an angle of 30 deg or less with respect to the vector A.

(6) (b) $C - D = (2, -3) = A - B$, hence, equivalent. (c) The vector from A' to B' where $A' = (3 + a, -2 + b)$ and $B' = (1 + a, 1 + b)$ and a and b are specific constants will work.

(7) The vector from A' to B' where $A' = (1 + a, 2 + b, -5 + c)$ and $B' = (1 + a, 2 + b, 1 + c)$ and a, b and c are specific constants will work.

(8) (a) The line $y = x$ (b) The line through $(2, 1)$ parallel to the line $y = x$ (c) Same as (b) since $t(-3, -3)$ is another parametric description for the line $y = x$, (d) same as (b) since $(1, 0)$ lies on the line from (b).

(9) Let (x, y) be a point on the line. Then $x = 2 + t$, $y = 1 + t$ so $y - 1 = t = x - 2$ or $y = x - 1$. Conversely suppose that (x, y) satisfies $y = x - 1$. Let $t = x - 2$. Then $x = 2 + t$, $y = 1 + t$, showing that (x, y) is a point on the line.

(10) Let (x, y) be a point on the line. Then $x = c + te$, $y = d + tf$. If $e = 0$, then the line is described by $x = c$ and is vertical. If $e \neq 0$, then $t = (x - c)/e$ showing that $y = \frac{f}{e}x + d - \frac{cf}{e}$

(11) The answer is all of R^2. Note: Students find this surprisingly "non-obvious."

(12) (a) The line $y = x$ in the xy plane, (b) the line through $(1, 0, 1)$ parallel to the line in (a), (c) the plane through the origin containing the vectors $(1, 1, 0)$ and $(1, 0, 1)$. This plane can also be described as the plane which contains the line $y = x$ in the xy plane and the line $z = x$ in the xz plane. (d) the plane from (c), translated one unit to the right along the x axis.

1.1. VECTORS

(13) Letting $x = 0$ and $x = 1$, we find that $(0, 2)$ and $(1, 5)$ both lie on the line. Hence $(x, y) = (0, 2) + t((1, 5) - (0, 2)) = (0, 2) + t(1, 3)$ is one parametric description. For the proof note that $x = t$, $y = 2 + 3t = 2 + 3x$. Hence (x, y) satisfies the original equation. Conversely, if (x, y) satisfies the original equation, then $(x, y) = (0, 2) + t(1, 3)$ where $t = x$, proving that (x, y) is a point on the line.

(14) The points $(0, 2)$ and $(1, 5)$ both lie on the line. Hence $(0, 2) + t(1, 3)$ and $(1, 5) + t(1, 3)$ are two different parametric descriptions of the line. For the proof, let $t = s - 1$ in the second description. This shows that every point on the second line is also on the first. Conversely, replacing t by $s + 1$ shows that every point on the first is also on the second.

(15) $(2, -2, 1)t + (1, -1, 1)$ works. So does $(2, -2, 1)t + (3, -3, 2)$

(16) The point on the line with z coordinate 0 is $(-1, 1, 0)$. The vector $(-2, 2, -1)$ is parallel to the line. Hence one answer is $(-2, 2, -1)t + (-1, 1, 0)$.

(17) $(1, -2, 1)t + (2, 4, 5) = (2, 4, 4)s + (4, 4, 8)$ if and only if $t = 1$ and $s = -\frac{1}{2}$ which produces the point $(3, 2, 6)$.

(18) The point on the line with z coordinate 0 is $(-3, 14, 0)$. The vector $(-1, 2, -1)$ is parallel to the line. Hence one answer is $(-1, 2, -1)t + (-3, 14, 0)$.

(19) The point $(2, 4, 4)t + (4, 4, 8)$ on the first line is on the second line if there is a number s such that $(2, 4, 4)t + (4, 4, 8) = (1, 2, 2)s + (2, 0, 4)$. This yields the three equations $2t + 4 = s + 2$, $4t + 4 = 2s$, and $4t + 8 = 2s + 4$, each of which is true if and only if $s = 2t + 2$. In fact, if we replace s by $2t + 2$ in the second parametrization, we produce the first, showing that every point on the first line is on the second. Similarly, replacing t by $\frac{s}{2} - 1$ in the first parametrization produces the second showing that every point on the second line is also on the first.

(20) (a) The points $(3, 0, 1)$, $(0, 2, 1)$, and $(3, -2, 0)$ all work. (b) The dot product of each of the points from (a) with $(2, 3, -6)$ all equal 0. (c) The point (x, y, z) satisfies the equation if and only if $0 = 2x + 3y - 6z = (2, 3, -6) \cdot (x, y, z)$ which means that (x, y, z) is perpendicular to $(2, 3, -6)$. The set of all points perpendicular to a fixed vector is clearly a plane.

(21) (a) $A = (-1, 1, 0)$, $B = (-1, -1, -1)$, $C = (2, -1, 0)$ work (b) $(2, 3, -6) \cdot (A - B) = (2, 3, -6) \cdot (0, 2, 1) = 0$ and $(2, 3, -6) \cdot (C - B) = (2, 3, -6) \cdot (3, 0, 1) = 0$. (c) $(x, y, z) - B = (x + 1, y + 1, z + 1)$ which is perpendicular to $(2, 3, 6)$ if and only if $0 = (x + 1, y + 1, z + 1) \cdot (2, 3, -6) = 2x + 3y - 6z - 1$ so the set of solutions is the plane through B perpendicular to $(2, 3, -6)$.

(22) The geometric interpretation is that the sum of the squares of the four sides of a parallelogram equals the sum of the squares of the diagonals. For the proof in R^2, let $A = (x_1, y_1)$, $B = (x_2, y_2)$. Then

$$\begin{aligned} |A + B|^2 + |A - B|^2 &= (x_1 + x_2)^2 + (y_1 + y_2)^2 \\ &\quad + (x_1 - x_2)^2 + (y_1 - y_2)^2 \\ &= 2x_1^2 + 2y_1^2 + 2x_2^2 + 2y_2^2 = 2|A|^2 + 2|B|^2 \end{aligned}$$

1.2 The Vector Space of $m \times n$ Matrices

True-False Answers: 1. T, 2. F, 3. F, 4. T, 5. F, 6. T, 7. F, 8. T

(1) (a) $\begin{bmatrix} -1 & -4 & -7 & -10 \\ 1 & -2 & -5 & -8 \\ 3 & 0 & -3 & -6 \end{bmatrix}$

third row: $[3, 0, -3, -6]$, second column $\begin{bmatrix} -4 \\ -2 \\ 0 \end{bmatrix}$

(b) $\begin{bmatrix} 1 & 8 \\ 4 & 32 \\ 9 & 72 \end{bmatrix}$

third row: $[9, 72]$, second column $\begin{bmatrix} 8 \\ 32 \\ 72 \end{bmatrix}$

(c) $\begin{bmatrix} \frac{1}{2} & -\frac{1}{2} \\ -\frac{1}{2} & -\frac{1}{2} \\ -1 & 0 \end{bmatrix}$

third row: $[-1, 0]$, second column $\begin{bmatrix} -\frac{1}{2} \\ -\frac{1}{2} \\ 0 \end{bmatrix}$

1.2. THE VECTOR SPACE OF M × N MATRICES

(2)
$$(A+B)+C = \begin{bmatrix} 3 & 1 & 3 \\ 4 & 2 & -2 \\ 4 & 3 & 3 \\ 2 & 4 & -1 \end{bmatrix} + \begin{bmatrix} 3 & 1 & 3 \\ 4 & 2 & -2 \\ 4 & 3 & 3 \\ 2 & 4 & -1 \end{bmatrix} = \begin{bmatrix} 6 & 2 & 6 \\ 8 & 4 & -4 \\ 8 & 6 & 6 \\ 4 & 8 & -2 \end{bmatrix}$$

$$A+(B+C) = A = \begin{bmatrix} 1 & 1 & 2 \\ 0 & 1 & -2 \\ 2 & 0 & 1 \\ 3 & 2 & 1 \end{bmatrix} + \begin{bmatrix} 5 & 1 & 4 \\ 8 & 3 & -2 \\ 6 & 6 & 5 \\ 1 & 6 & -3 \end{bmatrix} = \begin{bmatrix} 6 & 2 & 6 \\ 8 & 4 & -4 \\ 8 & 6 & 6 \\ 4 & 8 & -2 \end{bmatrix}$$

(3) $C = A + B$.

(4)
$$A+B+C = \begin{bmatrix} 6 & 2 & 6 \\ 8 & 4 & -4 \\ 8 & 6 & 6 \\ 4 & 8 & -2 \end{bmatrix}$$

(5) (a) $[1,1,4] = [1,1,2] + 2[0,0,1]$

   (b)
$$\begin{bmatrix} 1 & 2 \\ 0 & 0 \end{bmatrix} = 0 \begin{bmatrix} 0 & 0 \\ 1 & 0 \end{bmatrix} + \begin{bmatrix} 1 & 0 \\ 0 & 0 \end{bmatrix} + 2 \begin{bmatrix} 0 & 1 \\ 0 & 0 \end{bmatrix}$$

   (c)
$$\begin{bmatrix} 1 \\ 2 \\ 3 \end{bmatrix} = \begin{bmatrix} 4 \\ 5 \\ 6 \end{bmatrix} - \begin{bmatrix} 3 \\ 3 \\ 3 \end{bmatrix} + 0 \begin{bmatrix} 9 \\ 12 \\ 15 \end{bmatrix}$$

   (d)
$$\begin{bmatrix} 1 & 1 \\ 0 & 1 \end{bmatrix} = 0 \begin{bmatrix} 1 & 1 \\ 2 & 3 \end{bmatrix} + \begin{bmatrix} 1 & 2 \\ 0 & 0 \end{bmatrix} + \begin{bmatrix} 0 & -1 \\ 0 & 1 \end{bmatrix}$$

(6) A_1 cannot be a linear combination of A_2, A_3 and A_4 because all such linear combinations will have their $(2,1)$ entry equal to zero.

(7) Each vector has a non-zero entry in the positions where the other two vectors have zeros.

(8) $[1,-1,0], [1,0,0], [2,-2,0]$, and $[4,-1,0]$ all belong to the span. $[1,1,1]$ does not because its last entry is nonzero.

(9) (a) $-2X+Y = [1,1,4]$ works (b) Let $[x,y,z] = aX+bY = [-a-b, a+3b, -a+2b]$ then
$$5x + 3y - 2z = 5(-a-b) + 3(a+3b) - 2(-a+2b) = 0$$

(c) Any point $[x,y,z]$ which does not solve the equation $5x+3y-2z = 0$ will work–e.g. $[1,1,1]$.

(10) No. For $aX+bY$ to have the second and third entries positive, both a and b must be negative, in which case the first entry will be negative.

(11) For the first part, one could use X, Y, Z and $X+Y+Z = \begin{bmatrix} 4 & 2 \\ 0 & 4 \end{bmatrix}$.

For the second part, any matrix whose $(2,1)$ entry is nonzero will not be in the span.

(12) (a) The line containing the origin and the point $(1,2)$.
(b) All of R^2.
(c) No. From part (b) it appears that if A and B are independent elements of $M(1,2)$ then any other element of $M(1,2)$ will be a linear combination of them.
(d) The plane containing the two given vectors.
(e) $[1,2,1] = [1,1,0]+[0,1,1]$. Hence both $[1,2,1]$ and $[0,1,1]$ belong to the plane from part (e) and the planes in parts (d) and (e) are the same.
(f) The span of these vectors is the line through the origin containing each of them. Two independent vectors will span a plane but two dependent vectors will span a line.

(13) Let V and W be elements of the span. Then $V = aX + bY$ and $W = cX + dY$. Hence $xV + yW = (xa + yc)X + (xb + yd)Y$. This proves that $xV + yW$ is an element of the span of X and Y.

(14) The first row is twice the second plus twice the third.

(15) $3\begin{bmatrix} 4 \\ 1 \\ 1 \end{bmatrix} - \begin{bmatrix} 6 \\ 2 \\ 1 \end{bmatrix} = \begin{bmatrix} 6 \\ 1 \\ 2 \end{bmatrix}$

(16) No. If, say, the second row is a multiple of the first, then the matrix has the form
$$\begin{bmatrix} a & b \\ ca & cb \end{bmatrix}$$

1.2. THE VECTOR SPACE OF $M \times N$ MATRICES

If $a \neq 0$, then the second column is $\frac{b}{a}$ times the first. Similarly, if $b \neq 0$ then the first column is $\frac{a}{b}$ times the second. If both a and b are zero, then the columns are both zero, making the set of columns dependent. A similar argument applies if the second row is a multiple of the first in the original matrix.

(17) All sorts of answers here are possible. Be sure to request that the students state the dependency relation.

(18) $\{A, B, D\}$ is dependent: $D = 2A+B+3(A-B) = 5A-2B$. $\{A, C, D\}$ is dependent: $D = 2A + (C - A) + 3C = A + 4C$. Nothing can be concluded about the dependence of A and B.

(19) (a) $\sinh x = \frac{e^x - e^{-x}}{2} = \frac{1}{4}(2e^x) - \frac{1}{6}(3e^{-x})$.
   (b) $\cosh x - \sinh x = e^{-x}$.
   (c) $\cos(2x) = \cos^2 x - \sin^2 x$
   (d) $\cos(2x) = \cos^2 x - (1 - \cos^2 x) = 2\cos^2 x - 1$
   (e) $\sin x = \frac{\sqrt{2}}{2}(\sin(x + \frac{\pi}{4}) - \cos(x + \frac{\pi}{4}))$.
   (f) $(x + 3)^2 = x^2 + 6x + 9$
   (g) $x^2 + 3x + 3 = 3(x + 1) + \frac{1}{2}(2x^2)$
   (h) $\ln \frac{(x^2+1)^3}{x^4+7} = 6 \ln \sqrt{x^2 + 1} - \ln(x^4 + 7)$

(20) The span is the set of all polynomial functions of degree less than or equal to two.

(21) (a) Let $B = \begin{bmatrix} x & y \\ z & w \end{bmatrix}$. Then

$$A + B = \begin{bmatrix} x + a & y + b \\ z + c & w + d \end{bmatrix} = \begin{bmatrix} x & y \\ z & w \end{bmatrix}$$

Hence $x + a = x$, $y + b = y$, $z + c = c$, and $w + d = d$ which imply that $x = y = z = w = 0$. Hence $B = \mathbf{0}$.

(b) Let notation be as in (a). Then $A + B = \mathbf{0}$ if and only if

$$\begin{bmatrix} x + a & y + b \\ z + c & w + d \end{bmatrix} = \begin{bmatrix} 0 & 0 \\ 0 & 0 \end{bmatrix}$$

which implies that $x + a = 0$, $y + b = 0$, $z + c = 0$, $w + d = 0$. The conclusion follows.

(22) Let $A = [a_{ij}]$ and $B = [b_{ij}]$ be $m \times n$ matrices and let k and l be scalars. For (b), using the commutative law for numbers, $A + B = [a_{ij} + b_{ij}] = [b_{ij} + a_{ij}] = B + A$. For (e), let $-A = [-a_{ij}]$. Then $A + (-A) = [a_{ij} - a_{ij}] = \mathbf{0}$. For (h) $k(A + B) = k[a_{ij} + b_{ij}] = [k(a_{ij} + b_{ij})] = [ka_{ij} + kb_{ij}] = [ka_{ij}] + [kb_{ij}] = kA = kB$. Finally, for (i), $(k + l)A = [(k + l)a_{ij}] = [ka_{ij}] + [la_{ij}] = kA + lA$.

(23) In order, we used properties (c), (e), (d), (f), (g), (j).

(24) From Proposition 1,

$$\begin{aligned} -(2X + 3Y) &= (-1)(2X + 3Y) \\ &= (-1)(2X) + (-1)(3Y) = (-2)X + (-3)Y \end{aligned}$$

(In the last step, we used vector space property (g).)

(25)

$$\begin{aligned} -(aX) + (aX + (bY + cZ)) &= -(aX) + \mathbf{0} \\ (-(aX) + aX) + (bY + cZ) &= -(aX) \\ \mathbf{0} + (bY + cZ) &= -1(aX) \\ bY + cZ &= (-a)X \\ (-a)^{-1}(bY + cZ) &= (-a)^{-1}((-a)X) \\ (-a)^{-1}(bY) + (-a)^{-1}(cZ) &= ((-a)^{-1}(-a))X = 1X \\ \left(-\frac{b}{a}\right)Y + \left(-\frac{c}{a}\right)Z &= X \end{aligned}$$

The properties used were (in order): (a) and (e); (c) and (d); (b), (e), and Proposition 1; (b), (d) and (g); (f); (h) and (g); (g) and (j).

(26)

$$\begin{aligned} -A + (A + B) &= -A + \mathbf{0} \\ (-A + A) + B &= -A \\ \mathbf{0} + B &= -A \\ B &= -A \end{aligned}$$

The properties used were (in order): (a) and (e); (c) and (d); (b) and (e); (b) and (d).

(27) $A + (-1)A = 1A + (-1)A = (1 + (-1))A = 0A = \mathbf{0}$.

1.3 Systems of Linear Equations

True-False Answers: 1. T, 2. F, 3. F, 4. F, 5. T.

(1) X is a solution. Y is not.

(2) $aZ + bX = [a + b, a + b, a + b, 2a + b]$ which satisfies both equations, regardless of a and b.

(3) $aX + bY = [a + b, a + 2b, a - b, a + b]$. Substituting this into the first equation yields 0 and into the second yields $7b$. Hence, both equations will be satisfied if and only if $b = 0$.

(4) (a) Solution: the line $[11/7, 6/7, 1/7, 0]^t + w[-10/7, -1/7, 23/14, 1]$. (b) Solution: the line $[11, -6, 2, 0]^t + x_4[-1, 1, -1, 1]^t$.

(5) (Roman numerals are equation numbers)

 (a) Solution: $[-\frac{59}{9}, \frac{20}{9}, \frac{8}{9}]^t$, spanning: **0**, translation: $[-\frac{59}{9}, \frac{20}{9}, \frac{8}{9}]^t$.

 (b) Solution: the line $[1, 0, 0]^t + t[-\frac{17}{2}, \frac{5}{2}, 1]^t$, $2I + II = III$

 (c) Inconsistent: $2I+II$ contradicts III.

 (d) Solution: the line $[3/2, -1/2, 0]^t + z[-9/10, 1/10, 1]^t$, III=I+2II.

 (e) Solution: the line $[0, -1/3, 0]^t + z[-1, 01]^t$

 (f) Solution: $[6/25, -9/25, 7/5]^t$, spanning: **0**, translation: $[6/25, -9/25, 7/5]^t$.

 (g) Solution: the plane $[\frac{5}{4}, -\frac{1}{4}, 0, 0]^t + r[-\frac{3}{4}, -\frac{1}{4}, 1, 0]^t + s[-1, 0, 0, 1]^t$, spanning: $[\frac{5}{4}, -\frac{1}{4}, 0, 0]^t$, translation: $[-\frac{3}{4}, -\frac{1}{4}, 1, 0]^t$ and $[-1, 0, 0, 1]^t$, IV=I+2II, III=4I-II.

 (h) inconsistent, 2II+I contradicts IV.

(6) Letting t be 2 and 4 in the solution, produces respectively $X = [-16, 5, 2]$ and $Y = [-33, 10, 4]$. Hence $(X+Y)/2 = [-49/2, 15/2, 3]^t$, which satisfies each equation in the system, while $X + Y = [-49, 15, 6]^t$ satisfies none of the equations.

(7) (a) Choose any system of four independent equations in five unknowns. Then let the fifth equation be a linear combination of the other four.(b) Choose any system of three independent equations in five unknowns. Then let the fourth and fifth equations be a linear combinations of the first three.(c) Choose any system of two independent equations in three unknowns. Then let the third, fourth, and fifth equations be a linear combinations of the first two. (d) All five equations must be a multiple of the first equation.

(8) (a) For rank r, begin system with r independent equations. Then let the remaining equations be linear combinations of them. (b) You could simply let the first equation contradict the second.

Remark: To force the students to create a non-trivial example, you might request an inconsistent system of three equations in five unknowns which becomes consistent if any one equation is deleted.

(9) A point (x, y) solves the system if and only if it lies on both lines. Since the lines are parallel, there is no solution to the system.

(10) From the graph, all three lines must have the same y intercept which forces $a = 1$.

(11) False: The two equations could describe the same plane, in which case the solution set would be a plane. The two equations could also describe two parallel planes, in which case there are no solutions.

(12) The solution set will be a line as long as the planes described by the equations all intersect in the same line.

1.4 Gaussian Elimination

True-False Answers: 1. T, 2. F, 3. T, 4. T.

(1) (a)
$$\begin{bmatrix} 1 & 0 & -\frac{1}{2} & 0 & 5 \\ 0 & 1 & 1 & 0 & -1 \\ 0 & 0 & 0 & 1 & 2 \end{bmatrix}$$

(b) and (c)
$$\begin{bmatrix} 1 & 0 & 0 & 0 \\ 0 & 1 & 0 & 0 \\ 0 & 0 & 1 & 0 \\ 0 & 0 & 0 & 1 \end{bmatrix}$$

(2) (a) $[5 + 1/2s, -1 - s, s, 2]^t$ (b) and (c) are inconsistent.

(3) (a) neither (b) echelon (c) neither

(4) (a) Solution: $[0, -4 - s, -5/2 + s, 3/2 - 1/2s, s]^t$. Reduced form:
$$\begin{bmatrix} 1 & 0 & 0 & 0 & 0 & 0 \\ 0 & 1 & 0 & 0 & 1 & -4 \\ 0 & 0 & 1 & 0 & -1 & -5/2 \\ 0 & 0 & 0 & 1 & 1/2 & 3/2 \end{bmatrix}$$

1.4. GAUSSIAN ELIMINATION

(b) Solution: Inconsistent. Reduced form:
$$\begin{bmatrix} 0 & 1 & 0 & -2 & 0 \\ 0 & 0 & 1 & 2 & 0 \\ 0 & 0 & 0 & 0 & 1 \end{bmatrix}$$

(c) Solution: $[-3/2, 5/2, -1/2, 1/2]^t$. Reduced form:
$$\begin{bmatrix} 1 & 0 & 0 & 0 & -3/2 \\ 0 & 1 & 0 & 0 & 5/2 \\ 0 & 0 & 1 & 0 & -1/2 \\ 0 & 0 & 0 & 1 & 1/2 \end{bmatrix}$$

(5) (a)
$$\begin{bmatrix} 1 & 0 & 1 & 0 & 3 \\ 0 & 1 & -1 & 0 & 1 \\ 0 & 0 & 0 & 1 & 0 \end{bmatrix}$$

(b)
$$\begin{bmatrix} 1 & 0 & 0 & 0 & -1 \\ 0 & 1 & 0 & 0 & 1 \\ 0 & 0 & 1 & 0 & 0 \\ 0 & 0 & 0 & 1 & 1 \end{bmatrix}$$

(c)
$$\begin{bmatrix} 1 & 0 & 10/3 \\ 0 & 1 & 1/3 \end{bmatrix}$$

(d)
$$\begin{bmatrix} 1 & 1/2 & 0 & 5 & 0 & -1/2 \\ 0 & 0 & 1 & -2 & 0 & 0 \\ 0 & 0 & 0 & 0 & 1 & 1 \end{bmatrix}$$

(e) and (f)
$$\begin{bmatrix} 1 & 0 \\ 0 & 1 \end{bmatrix}$$

(6) From the reduced form of A (below) the system is solvable if and only if $c = a + 2b$.
$$\begin{bmatrix} 3 & 2 & -1 & a \\ 0 & 1/3 & 7/3 & b - a/3 \\ 0 & 0 & 0 & c - a - 2b \end{bmatrix}$$

(7) (a) and (b) The equation $B = x_1 X_1 + x_2 X_2 + x_3 X_3$ yields the system with the following augmented matrix and echelon form which represents a consistent system, showing that B is in the span.

$$\begin{bmatrix} 1 & 1 & 1 & a \\ 0 & 2 & 1 & b \\ -1 & 1 & 1 & c \end{bmatrix} \quad \begin{bmatrix} 1 & 1 & 1 & a \\ 0 & 2 & 1 & b \\ 0 & 0 & 1 & c+a-b \end{bmatrix}$$

(c) The equation $B = x_1 X_1 + x_2 X_2 + x_3 X_3$ yields the system with the following augmented matrix and echelon form which represents an inconsistent system, showing that B is not in the span.

$$\begin{bmatrix} 1 & 1 & 3 & 3 \\ 0 & 2 & 4 & 2 \\ -1 & 1 & 1 & 1 \end{bmatrix} \quad \begin{bmatrix} 1 & 1 & 3 & 3 \\ 0 & 2 & 4 & 2 \\ 0 & 0 & 0 & 2 \end{bmatrix}$$

(8) The vector equation $[a, b]^t = xX + yY$ yields the system with augmented matrix on the left. Reduction produces the matrix on the right which represents a consistent system, showing that X and Y span R^2.

$$\begin{bmatrix} 1 & 1 & a \\ 2 & -2 & b \end{bmatrix} \quad \begin{bmatrix} 1 & 1 & a \\ 0 & -4 & b-2a \end{bmatrix}$$

(9) The student can just pick a vector at random. The chances are that it won't be in the span. More explicitly, if $B = [a, b, c]^t$, then the equation $B = x_1 X_1 + x_2 X_2 + x_3 X_3$ yields the system with augmented matrix on the left and reduced form on the right. As long as the student doesn't choose a vector such that $a - c + 2b = 0$, it will not be in the span.

$$\begin{bmatrix} 5 & 4 & 3 & c \\ 1 & 1 & 2 & b \\ 3 & 2 & -1 & a \end{bmatrix} \quad \begin{bmatrix} 5 & 4 & 3 & c \\ 0 & 1/5 & 7/5 & b-(1/5)c \\ 0 & 0 & 0 & a-c+2b \end{bmatrix}$$

(10) The reduced form of the system will have three non-zero equations in five variables. Therefore, two of the variables will be free and can be set arbitrarily.

(11)
$$\begin{bmatrix} 1 & 0 \\ 0 & 1 \end{bmatrix} \quad \begin{bmatrix} 1 & a \\ 0 & 0 \end{bmatrix} \quad \begin{bmatrix} 0 & 1 \\ 0 & 0 \end{bmatrix} \quad \begin{bmatrix} 0 & 0 \\ 0 & 0 \end{bmatrix}$$

1.4. GAUSSIAN ELIMINATION

(12) The identity matrix I, the zero matrix $\mathbf{0}$, and

$$\begin{bmatrix} 1 & 0 & a \\ 0 & 1 & b \\ 0 & 0 & 0 \end{bmatrix} \begin{bmatrix} 1 & a & 0 \\ 0 & 0 & 1 \\ 0 & 0 & 0 \end{bmatrix} \begin{bmatrix} 1 & a & b \\ 0 & 0 & 0 \\ 0 & 0 & 0 \end{bmatrix}$$

$$\begin{bmatrix} 0 & 1 & 0 \\ 0 & 0 & 1 \\ 0 & 0 & 0 \end{bmatrix} \begin{bmatrix} 0 & 1 & a \\ 0 & 0 & 0 \\ 0 & 0 & 0 \end{bmatrix} \begin{bmatrix} 0 & 0 & 1 \\ 0 & 0 & 0 \\ 0 & 0 & 0 \end{bmatrix}$$

(13) The system is consistent since $\mathbf{0}$ is a solution. Hence, the more unknowns theorem says that there are an infinite number of solutions.

(14) There can be at most 3 non-zero rows.

(15) (a) No, the system is not solvable for all values of a, b and c. (b) The identity matrix (c) No, there will be a free variable. (d) No, the reduced form has a row of zeros, showing that there are constants for which there is no solution. (e) No. Either there will be free variables, proving that there is not a unique solution, or there will be a row of zeros in the reduced form, showing that there will be instances where the corresponding system is not solvable.

(16) If $a = 0$, then $bc = 0$ so b or c is zero. In either case, the theorem is clearly true. If $a \neq 0$, then $d = bc/a$ and the last row of A is $[c, bc/a] = \frac{c}{a}[a,b]$, proving dependence.

(17) $x[1, 0, 1, -2]^t + y[0, 1, -1, 1]^t$

(18) The system has a solution if and only if $x = 2y$.

1.4.1 Network Flow

(1) General solution: $[20 - w, 30 + w, 50 - w, w]^t$.

 (a) The greatest traffic flow was on West St. where the traffic varied from 44 to 42 cars per min.

 (b) **THIS EXERCISE IS UNWORKABLE AS STATED. THE TRAFFIC IS NEVER HEAVIEST ON WEST STREET.**

(2) The system is

$$\begin{aligned} -x + y & = 90 \\ y - z & = 20 \end{aligned}$$

CHAPTER 1. SYSTEMS OF LINEAR EQUATIONS

$$\begin{aligned} -z + w &= 30 \\ w - v &= 60 \\ -x + v &= 40 \end{aligned}$$

(3)
$$\begin{aligned} -x   + s &= 100 \\ x - u &= 50 \\ y + t - s &= 50 \\ y - u + v &= 300 \\ z + t &= 350 \\ z + v &= 450 \end{aligned}$$

The general solution is $[-100 + s, 50 - t + s, 350 - t, -150 + s, 100 + t, t, s]^t$. The smallest allowed value of s is $s = 150$, in which case the solution becomes $[50, 200 - t, 350 - t, 0, 100 + t, t, 150]^t$ where $0 \le t \le 200$.

1.5 Column Space and Nullspace

1. T, 2. F, 3. T, 4. T, 5. T, 6. F, 7. F, 8. T, 9. F.

(1) (a) $[0, 5, -11]^t$ (b) $[7, 10, 7, 5]^t$ (c) $[x_1 + 2x_2 + 3x_3, 4x_1 + 5x_2 + 6x_3]^t$

(2)

(4a) $A = \begin{bmatrix} 1 & -1 & 2 & -2 \\ 2 & 1 & 0 & 3 \\ 2 & 3 & 2 & 0 \end{bmatrix}$ $B = \begin{bmatrix} 1 \\ 4 \\ 6 \end{bmatrix}$

(4b) $A = \begin{bmatrix} 1 & 2 & 1 & 0 \\ 0 & 1 & 4 & 3 \\ 0 & 0 & 2 & 2 \end{bmatrix}$ $B = \begin{bmatrix} 1 \\ 2 \\ 4 \end{bmatrix}$

(5f) $A = \begin{bmatrix} 2 & 3 & -1 \\ 1 & -1 & 1 \\ 2 & 3 & 4 \end{bmatrix}$ $B = \begin{bmatrix} -2 \\ 2 \\ 5 \end{bmatrix}$

(5g) $A = \begin{bmatrix} 1 & 1 & 1 & 1 \\ 2 & -2 & 1 & 2 \\ 2 & 6 & 3 & 2 \\ 5 & -3 & 3 & 5 \end{bmatrix}$ $B = \begin{bmatrix} 1 \\ 3 \\ 1 \\ 7 \end{bmatrix}$

1.5. COLUMN SPACE AND NULLSPACE

(3) Any equation of the form
(a)
$$\begin{aligned} x \phantom{{}+2y} - 3z + 2w &= a \\ 2x - 2y + z + w &= b \\ 3x + 2y - 2z - 3w &= c \end{aligned}$$

(b)
$$\begin{aligned} -5x + 17y &= a \\ 4x + 2y &= b \\ 3x + y &= c \\ 5x - 5y &= d \end{aligned}$$

(c)
$$\begin{aligned} x + 2y + 3z &= a \\ 4x + 5y + 6z &= b \end{aligned}$$

(4) The nullspace is spanned by : (a) $\{[4/7, 3/2, 6/7, 1]^t\}$, (b) $\{[0,0]^t\}$ (c) $\{[1, -2, 1]^t\}$

(5) The nullspace is spanned by : (a) $\{[-1, 1, 1, 0, 0]^t, [-3, -1, 0, 0, 1]^t\}$ (b) $\{[1, -1, 0, -1, 1]^t\}$ (c)$\{[-10, -1, 3]^t\}$
(d) $\{[1/2, 0, 0, 0, -1, 1]^t, [-5, 0, 2, 1, 0, 0]^t, [-1/2, 1, 0, 0, 0, 0]^t\}$ (e) $\{[0,0]\}$
(f) $\{[0,0]\}$

(6) Let the columns of B be scalar multiples of $[1, 2]^t$

(7) Let the columns of B be scalar multiples of $[1, 2, 3]^t$

(8) Let the columns of B be any four vectors which span the same space as the given vectors.

(9) See the answer for Exercise 13, Section 1.4.

(10) The nullspace is the set of vectors of the following form. This exercise demonstrates the translation theorem.

$$s \begin{bmatrix} 1 \\ 1 \\ 0 \\ 0 \\ 0 \\ 0 \end{bmatrix} + t \begin{bmatrix} -2 \\ 0 \\ -1 \\ 1 \\ 0 \\ 0 \end{bmatrix} + u \begin{bmatrix} -1 \\ 0 \\ 1 \\ 0 \\ 1 \\ 0 \end{bmatrix}$$

(11) One checks by substitution that $[-1, 1, 2, 1, 1, 1]^t$ is a particular solution to the non-homogeneous system. It follows from the answer to Exercise 2 in Section 1.4 and the translation theorem that the vectors on the right in Exercise 11 span the nullspace of A. Thus, the translation theorem says that the expression in Exercise 11 does represent the general solution.

(12) One checks by substitution that $[-1, 1, 2, 1, 1, 1]^t$ is a particular solution to the non-homogeneous system. Note that

$$\begin{bmatrix} 1 \\ 1 \\ 0 \\ 0 \\ 0 \\ 0 \end{bmatrix} + \begin{bmatrix} -2 \\ 0 \\ -1 \\ 1 \\ 0 \\ 0 \end{bmatrix} = \begin{bmatrix} -1 \\ 1 \\ -1 \\ 1 \\ 0 \\ 0 \end{bmatrix}$$

Thus, the three vectors on the right in Exercise 12 span the same subspace as the three vectors on the right in Exercise 11 which is the nullspace of A. Thus, it follows from the translation theorem that the expression in Exercise 12 is the general solution to the system.

(13) True. $Y_1 = 2X_1 + 2X_2$, $Y_2 = X_1 - X_2$. Hence, Y_1 and Y_2 belong to the span of the X_i. Since spans are subspaces, the span of the Y_i is contained in the span of the X_i. Conversely, $X_1 = \frac{1}{4}Y_1 + \frac{1}{2}Y_2$ and $X_2 = \frac{1}{4}Y_1 - \frac{1}{2}Y_2$ showing that the span of the X_i is contained in the span of the Y_i. Hence, the spans are equal.

(14) False. Note that $Y_1 = X_1 + X_2$, $Y_2 = X_1 - X_2$ and $Y_3 = X_2$ showing that the span of the Y_i is contained in the span of the X_i. However X_3 is not a linear combination of the Y_i since all of the Y_i have their last two entries equal and the last two entries of X_3 are unequal.

(15) (a) Let the equation be $ax + by + cz = d$. Since $\mathbf{0}$ belongs to the span, the zero vector solves the equation, showing that $d = 0$. Substituting $[1, 2, 1]^t$ and $[1, 0, -3]^t$ into the equation yields the system

$$\begin{aligned} a + 2b + c &= 0 \\ a - 3c &= 0 \end{aligned}$$

One solution is $c = 1$, $a = 3$, $b = -2$. (b) Let each equation be a multiple of the one from (a).

1.5. COLUMN SPACE AND NULLSPACE

(16) Let the equation be $ax + by + cz = d$. From the translation theorem, $[1, 2, 1]^t$ and $[1, 0, -3]^t$ must span the solution set for the homogeneous equation. From Exercise 15, one such equation is $3x - 2y + z = 0$. The vector $[1, 1, 1]^t$ is a particular solution to $3x - 2y + z = 2$. Any system in which each equation is a multiple of this equation would have the desired solution set.

(17) No, the two answers are not consistent. If the answers were consistent, then the difference of any two solutions to the system would be a solution to the homogeneous system which, form Group I's, answer is spanned by $[-3, 1, 1]^t$ and $[-1, 0, 1]^t$. Thus, there should exist s and t such that the following equation is true. The corresponding system is, however, inconsistent.

$$[1, 0, 0]^t - [1, -1, 1]^t = s[-3, 1, 1]^t + t[-1, 0, 1]^t$$

(18) For all scalars a, b, c, and d,

$$aX + bY + cZ + dW = (a + 3c)X + (b - 2d)Y$$

showing that the span of X, Y, Z, and W is contained in the span of X and Y. Conversely,

$$aX + bY = aX + bY + 0Z + 0W$$

showing the equality of the spans.

(19) (a) If W belongs to span $\{X, Y, Z\}$ then $W = aX + bY + cZ = aX + bY + c(2X + 3Y) = (a + 2c)X + (b + 3c)Y$ which belongs to span $\{X, Y\}$. Conversely, if W belongs to span $\{X, Y\}$, then $W = aX + bY = aX + bY + 0Z$ which belongs to span $\{X, Y, Z\}$. Thus, the two sets have the same elements and are therefore equal.

(20) Let $\mathcal{W}$ satisfy the subspace properties. Then $\mathcal{W}$ is non-empty since it contains the zero vector. If X and Y belong to $\mathcal{W}$ and a and b are scalars, then aX and bY both belong to $\mathcal{W}$. Hence $aX + bY$ also belongs to $\mathcal{W}$ showing that $\mathcal{W}$ is closed under linear combinations.

(22) Let X and Y be elements of $\mathcal{W}$. Then

$$X = \begin{bmatrix} a & b \\ c & d \end{bmatrix} \quad Y = \begin{bmatrix} a' & b' \\ c' & d' \end{bmatrix}$$

where $a + b + c + d = 0 = a' + b' + c' + d'$. Clearly, $\mathcal{W}$ contains the zero vector. If X is as above and k is a scalar, then, $0 = k(a + b +$

$c+d) = ka+kb+kc+kd$ which is equivalent with kX belonging to $\mathcal{W}$. Similarly,

$$0 = (a+b+c+d)+(a'+b'+c'+d') = (a+a')+(b+b')+(c+c')+(d+d')$$

which is equivalent with $X+Y$ belonging to $\mathcal{W}$. This finishes the proof.

(23) $\mathcal{W}$ is the first quadrant in R^2. No: $\mathcal{W}$ is not closed under scalar multiplication.

(24) $\mathcal{W}$ is not a subspace because it is not closed under scalar multiplication.

(25) Only lines through the origin form subspaces.

(26) If the first entry of either X or Y is zero, then $X+Y$ will belong to $\mathcal{W}$.

(27) Start by noting that the general upper triangular matrix is

$$A = \begin{bmatrix} a & b & c \\ 0 & d & e \\ 0 & 0 & f \end{bmatrix}$$

and then proceed as in Exercise 21a, part (iii).

(28) $aA+bB$ will be unipotent if and only if $a+b=1$. Hence the set of unipotent matrices is not closed under linear combinations and not a subspace.

(29) $aX+bY$ is a solution if and only if and only if $a+b=1$.

(30) The zero element belongs to $\mathcal{S} \cap \mathcal{T}$. If X and Y belong to $\mathcal{S} \cap \mathcal{T}$, then X and Y both belong to $\mathcal{S}$ so $aX+bY$ belongs to $\mathcal{S}$ for any scalars a and b. Similarly, $aX+bY$ belongs to $\mathcal{T}$; hence to $\mathcal{S} \cap \mathcal{T}$. Thus, $\mathcal{S} \cup \mathcal{T}$ is closed under linear combinations and is a subspace.

(31) $\mathcal{S} \cup \mathcal{T}$ is a subspace only if either $\mathcal{S} \subset \mathcal{T}$ or $\mathcal{T} \subset \mathcal{S}$. For the proof, suppose that $\mathcal{S}$ is not contained in $\mathcal{T}$. Then $\mathcal{S}$ contains an element S which is not in $\mathcal{T}$. Then for all T in $\mathcal{T}$, $U = S+T$ must be in $\mathcal{S} \cup \mathcal{T}$. But U cannot be in $\mathcal{T}$ since $S = T - U$ and S is not in $\mathcal{T}$. Thus, U belongs to $\mathcal{S}$, proving that $T = U - S$ belongs to $\mathcal{S}$. Hence, $\mathcal{T} \subset \mathcal{S}$.

1.5. COLUMN SPACE AND NULLSPACE

1.5.1 Predator-Prey Problems

(3) The maximum mountain lion population is around 14×10^2. The maximum rabbit population is around 11×10^2. The model predicts eventual extinction of both populations.

(4) 41.4×10^2. This rate of growth is not reasonable.

(5) The graph is a spiral which gets out of the first quadrant.

(6) (a) Formula (2) may be written

$$R_{n+1} = (1.1)(R_n - \frac{0.1}{1.1} M_n)$$

If we assume that the reproductive cycle of the rabbits is short, then the number of rabbits eaten prior to the reproductive period is

$$\frac{0.1}{1.1} M_n = \frac{1.0}{1.1} \times 10^2 \approx 90$$

The number of rabbits eaten does not depend on the number of rabbits in the population. Hence, the mountain lions can find all the rabbits they want.

Remark: In this discussion, we assumed that the reproduction occurs during a short period while the eating occurs throughout the year. Some such assumption is necessary to answer the question.

(b) Formula 1 suggests a food shortage for the mountain lions.

Chapter 2

Dimension

2.1 The Test for Linear Independence

True-False Answers: 1. F, 2. Only the second statement is true, 3. T, 4. T, 5. T, 6. F, 7. T, 8. T, 9. F.

(1) (a) independent (b) dependent: The third matrix is -2 times the first plus 3 times the second. (c) independent (d) dependent: The third matrix is 2 times the second (e) dependent: The third matrix is -3 times the first plus 4 times the second. (f) independent (g) independent.

(2) Let the rows of A be A_1, A_2, A_3 and A_4. We find from the test for independence that $A_3 = -3A_1 + 3A_2$ and $A_4 = 2A_1 + A_2$. Thus, the system has rank 2 and the last two equations may be omitted. We can actually disagreed any two of the equations. For example, A_1 is a linear combination of A_2 and A_4 so A_3 is also a linear combination of A_2 and A_4, allowing us to eliminate A_1 and A_3.

(3) Dependent: Let A_i be the columns of A. Then $A_3 = A_1 + A_2$ and $A_4 = A_1 - 2A_2$.

(4) The maximal number of independent columns equals the number of pivot variables which is the number of non-zero rows in the reduced form. We expect that this is also the number of independent rows. Thus, we expect that the maximal number of independent rows equals the maximal number of independent columns.

(5) Let the rows of A be A_1, A_2 and A_3. Then $xA_1 + yA_2 + zA_3 = \mathbf{0}$ yields the system

$$\begin{aligned} x &= 0 \\ ax &= 0 \\ y &= 0 \\ bx + dy &= 0 \\ z &= 0 \\ cx + ey + fz &= 0 \end{aligned}$$

The first, third and fifth equations show that $x = y = 0$, proving independence.

(6) Let the rows of A be A_1, A_2 and A_3. Then $xA_1 + yA_2 + zA_3 = \mathbf{0}$ yields the following system which is easily solved to yield $x = y = z = 0$.

2.1. THE TEST FOR LINEAR INDEPENDENCE

$$\begin{aligned} x &= 0 \\ ax &= 0 \\ bx + y &= 0 \\ cx + fy &= 0 \\ dx + gy + z &= 0 \\ ex + hy + zk &= 0 \end{aligned}$$

(7) In Exercise 5, $A_2 = aA_1$, $A_4 = bA_1 + dA_3$, $A_6 = cA_1 + eA_3 + fA_5$. In Exercise 6, $A_2 = aA_1$, $A_4 = (c - bf)A_1 + fA_3$, $A_6 = (e - bh - kd + kbg)A_1 + (h - gk)A_3 + kA_5$.

(8) The equation $xY_1 + yY_2 + zY_3 = 0$ is equivalent with

$$(x + y)X_1 + (2x + y + 7z)X_2, -xX_3 = 0$$

Since the X_i are independent, we see that $0 - x = 2x + y + 7z = x + y$ which implies that $x = y = z = 0$, proving independence.

(9) The dependency equations for the columns of A is

$$0 = x_1 A_1 + x_2 A_2 + \ldots + x_n A_n = AX$$

where $X = [x_1, x_2, \ldots, x_n]^t$. Thus, the columns are dependent if and only if the equation $AX = 0$ has a non-zero solution which is the same as saying that the nullspace is non-zero.

(10) Let the matrices be $[a_1, b_1]^t$, $[a_2, b_2]^t$ and $[a_3, b_3]^t$. The dependency equation for these vectors is equivalent with the system

$$\begin{aligned} a_1 x + b_1 y + c_1 z &= 0 \\ a_2 x + b_2 y + c_2 z &= 0 \end{aligned}$$

This system is consistent since the zero vector is a solution. Hence, it has a non-trivial solution from the more unknowns theorem.

(11) Let the matrices be A, B, C, D and E. Then the dependency equation $xA + yB + zC + wD + uE = 0$ results in a system of four homogeneous equations in five unknowns which has a non-zero solution from the same reasoning as in Exercise 10.

(12) 6. For the proof reason as in Exercise 11.

(13) (a) The dependency equation is $ae^x + be^{2x} + ce^{3x} = \mathbf{0}$. Differentiating twice, and setting $x = 0$ yields the system

$$\begin{aligned} a + b + c &= 0 \\ a + 2b + 3c &= 0 \\ a + 4b + 9c &= 0 \end{aligned}$$

which is easily solved to prove that $a = b = c = 0$, showing independence.

(b) Differentiating the dependency equation three times, and setting $x = 0$ yields the following system which is easily solved to prove that $a = b = c = 0$, showing independence.

$$\begin{aligned} a + b + c + d &= 0 \\ b + 2c + 3d &= 0 \\ b + 4c + 9d &= 0 \\ b + 8c + 27d &= 0 \end{aligned}$$

(c) Differentiating the dependency equation twice, and setting $x = 0$ yields the system $a = b = 2c = 0$, proving independence.

(d) Differentiating the dependency equation four times, and setting $x = 0$ yields the system $a = b = 2c = 6d = 24e = 0$, proving independence.

(e) Differentiating the dependency equation *two* times, and setting $x = 1$ yields the system $a + b = -a + b = 0$ which is easily solved to prove $a = b = 0$.

(f) Differentiating the dependency equation three times, and setting $x = 1$ yields the following system which is easily solved to prove $a = b = c = 0$, proving independence.

$$\begin{aligned} a + b + c &= 0 \\ -a + b + 3c &= 0 \\ 2a - b + 2c &= 0 \end{aligned}$$

2.2. DIMENSION

(14) The system of equations obtained by differentiating the dependency equation and substituting 0 is rank 2 and thus has non-zero solutions.

2.2 Dimension

True-False Answers   1. T, 2. T, 3. F, 4. F, 5. F, 6. F, 7. F, 8. T, 9. F.

(3) B can be any vector not of the form $[c, 2c]$ and C must be of this form.

(4) True. A_1 and A_2 span R^2 since they are independent and therefore A_1, A_2, and A_3 will also span.

(5) Yes: $Z = 2X_1 + 3X_2$.

(6) (a) Spanning set $\{[1, 2, 1, 1]^t, [1, 1, 1, 2]^t, [2, 3, 2, 3]\}$. The first two form a basis since the third is their sum. The second two are also a basis. The dimension is 2.

(b) Spanning set $\{[1, 2, 1]^t, [0, 1, 1]^t, [2, 3, 1]^t\}$. The first two form a basis since the third is twice the first minus the second. The second two are also a basis. The dimension is 2.

(c) Spanning set $\{[1, 2, 1]^t, [1, 1, 1]^t, [2, 3, 1]\}$. These vectors are also independent and form a basis. Another basis could be obtained by, say, replacing the first element by 2 times itself. The dimension is 3.

(7) $\mathcal{W}$ is a subspace because it is a span. A basis is

$$\begin{bmatrix} 1 & 1 \\ 2 & 1 \end{bmatrix} \quad \begin{bmatrix} 1 & 1 \\ 0 & 1 \end{bmatrix}$$

(9) The dimension is 6. For the proof, expand the general 2×3 matrix in similar manner to that done in Example 1.

(10) The dimension is mn. For the proof, the students should explain that the general element of $M(n, m)$ is a linear combination of the matrices E_{ij} which have a one in the (i, j) entry and all other entries equal zero. They should also explain that these elements are independent because each has a 1 in a position where all of the others have a 0.

(11) The dimension is 6. For the proof, expand the general 3×3 upper triangular matrix in similar manner to that done in Example 1.

30   CHAPTER 2. DIMENSION

(12) From the work done in Example 3 of Section 2.1, $X_1 - 2X_2 = X_3$ and $-2X_1 + X_2 = X_4$. Thus, X_1 and X_2 span the subspace in question. Since X_1 and X_2 are independent, the dimension is 2 and these two matrices form a basis.

(13) Let X_i denote the elements of each set, listed in the given order.(a) X_1 and X_2 are independent and $X_3 = 2X_1 + 3X_2$, $X_4 = 3X_1 + 10X_2$. (b) X_1, X_2, and X_3 are independent and $X_4 = -\frac{1}{2}X_1 + \frac{1}{2}X_2 + \frac{3}{2}X_3$.(c) These vectors are independent.

(14) (a) $\{[-1,0,1]^t, [-2,1,0]^t\}$ (b) $\{[-2/5, -1/5, 0, 1]^t, [-1, 0, 1, 0]^t\}$, (c)$\{[-4,0,0,1]^t, [-3,0,1,0]^t, [5,1,0,0]^t\}$.

(15) The augmented matrix for the system of equations corresponding to the dependency equation and an echelon form are, respectively,

$$A = \begin{bmatrix} 2 & 1 & 7 & 0 \\ -3 & 2 & -2 & 0 \\ 1 & 3 & 4 & 0 \end{bmatrix} \quad \begin{bmatrix} 2 & 1 & 7 & 0 \\ 0 & 7/2 & 17/2 & 0 \\ 0 & 0 & -39/7 & 0 \end{bmatrix}$$

Thus, the three vectors are independent and form a basis for R^3 since R^3 is three dimensional.

(16) We showed that the X_i were independent in Exercise 15. Without dimension theory, we must also show that they span. Let $Y = [a, b, c]^t$. The augmented matrix for the system corresponding to the equation $Y = xX_1 + yX_2 + zX_3$ and an echelon form are

$$B = \begin{bmatrix} 2 & 1 & 7 & a \\ -3 & 2 & -2 & b \\ 1 & 3 & 4 & c \end{bmatrix}$$

$$\begin{bmatrix} 2 & 1 & 7 & a \\ 0 & 7/2 & 17/2 & b + (3/2)a \\ 0 & 0 & -39/7 & c - (11/7)a - (5/7)b \end{bmatrix}$$

Thus, each of the desired variables may be found by back substitution, showing that the X_i span.

Remark: It is interesting to compare the amount of work that would have been saved by using dimension theory. From Theorem 2, since we are given three vectors *and we know that R^3 is three-dimensional*, it suffices to prove independence. For this, we need only reduce matrix A, not matrix B. This involves considerably less computation since B contains variables (a, b, and c) while A does not.

2.2. DIMENSION

Without dimension theory, we must, in principal, reduce both A and B. However, once we have reduced B, we have also reduced A, so we really only need to reduce B. Actually, with a little thought, we can get away with only reducing A, even if we choose not to use dimension theory. To explain this, note that what we really want to know is that the system of equations described by B is consistent, regardless of the values of a, b, and c. This will be true if the reduced form of B has no rows which express an equation of the form $0 = 1$. Looking at A, we can see that this is true since each row of A has a non-zero entry which is not in the last column. Thus, if we think carefully, the amount of work is the same with or without dimension theory. Dimension theory only saves some thought. These comments may, in fact, be used to provide a proof of Theorem 2 for the case of R^n.

(17) (a) Any vector which is independent of X_1 and X_2 will work. (b) There must exist a vector X_3 which is not in the span of X_1 and X_2 since these vectors span a two dimensional subspace. It follows as in the proof of Proposition 3 that then $\{X_1, X_2, X_3\}$ is independent and hence forms a basis. (c) X_4 may be any vector independent of X_1, X_2, and X_3. (d) As in (b) there exists a vector X_3 in R^4 independent of X_1 and X_2 and a vector X_4 independent of X_1, X_2, and X_3. (e) is proven using the same argument and induction.

(18) (a) It is easily proven that the X_i are independent; hence span the nullspace. (b) $[1,1,1,1]^t = \frac{1}{4}X_1 + \frac{1}{4}X_2 + \frac{1}{4}X_3$, showing that it does indeed belong to the nullspace. (c) $Y_1 = X_1$, $Y_2 = X_2$, $Y_3 = X_1 - X_2$, so the Y_i do belong to the nullspace. (d) The Y_i are dependent ($Y_3 = Y_1 - Y_2$); hence span only a two dimensional subspace of the nullspace.

(19) $\{A, B, C\}$ is dependent. $\{B, C\}$ need not be a basis of $\mathcal{W}$.

(20) It suffices to show that X and Y are independent. The dependency equation is $(2x + 3y)A + (3x - 5y)B = \mathbf{0}$. Since A and B are independent, this is true if and only if $0 = 2x + 3y = 3x - 5y$. This system is easily solved to prove that $x = y = 0$, finishing the exercise.

(23) The proof is similar to the argument done to prove Proposition 3.

(31) The dimension of P_n is $n + 1$.

2.2.1 Differential Equations

(1) (a) $\cosh x = \frac{1}{2}e^x + \frac{1}{2}e^{-x}$
(b) $\cos(2x) = \cos^2 x - \sin^2 x$

(c) $\sinh x + \cosh x = e^x$

(d) $(\frac{1}{2})1 - \frac{1}{2}\cos(2x) = \sin^2 x\}$

(e) $\ln(3x) = \ln x + (\ln 3)1$

(f) $x(x-2)^2 = 4x - 4x^2 + x^3$

(g) $2x^2 + 5x + 3 = 2(x-1)^2 + 9(x-1) + (10)1$

(4) (a) $\{e^{-x}, e^{-2x}\}$. (b) The roots of the characteristic polynomial are $-2 \pm 3i$. Hence, there are no solutions of the form e^{rx} where r is real. (c) The characteristic polynomial factors as $(r+1)^2(r-1)$ so the answer is $\{e^x, e^{-x}\}$. They do not span the solution space since the solution space is three dimensional. (d) The characteristic polynomial factors as $(r+1)^3$ yielding e^{-x} as a solution. The general solution is $y(x) = (a + bx + cx^2)e^{-x}$. (e) The general solution is $y(x) = a + bx + cx^2$.

(5) $y_1'(x) = e - 2e^{-2x}\cos(3x) - 3e^{-2x}\sin(3x)$ and $y_1''(x) = -5e^{-2x}\cos(3x) + 12e^{-2x}\sin(3x)$. $y_2'(x) - 2e^{-2x}\sin(3x) + 3e^{-2x}\cos(3x)$ and $y_2''(x) = -5e^{-2x}\sin(3x) - 12e^{-2x}\cos(3x)$.

(6) The general solution is $y(x) = (a+bx)e^{-x} + ce^x$. The equation $y(1) = 0$ is equivalent with $c = -e^{-2}(a+b)$ so

$$y(x) = a(e^{-x} - e^{-2}e^x) + b(xe^{-x} - e^{-2}e^x)$$

Thus, the space in question is the span of the functions $e^{-x} - e^{-2}e^x$ and $xe^{-x} - e^{-2}e^x$; hence a subspace. These functions are linearly independent since neither is scalar multiple of the other. Thus, the dimension is 2.

(7) The sum of any two solutions will solve the equation $y'' + 3y' + 2y = 2x^2$, not $y'' + 3y' + 2y = x^2$.

(8) General solution

$$y(x) = 7/4 - 3/2\,x + x^2/2 + C_1\,e^{-2x} + C_2\,e^{-x}$$

(9) The general solution is

$$y(x) = 1/10\,\cos(x) + 3/10\,\sin(x) + C_1\,e^{-2x} + C_2\,e^{-x}$$

(11) The set of all polynomials is a subspace. The set of polynomials with integer coefficients is not.

(13) The general solution is all polynomials of degree three or less.

2.3. APPLICATIONS TO SYSTEMS

(15) The dimension of P_n is $n+1$. The space of all polynomial functions is infinite dimensional.

(17) The students must show that the set of polynomials of the form $p_n(x) = (x-1)^n$ is independent.

(18) The space is infinite dimensional. To prove it, one can show that the polynomials of the form $p_n(x) = (x-1)^n(x-2)$ is independent. For this, the easiest proof is to note that if $\sum c_n P_n(x) = 0$, then (dividing by $(x-1)$) we see that $\sum c_n(x-1)^n = 0$ which proves that the c_n are zero since the set of functions of the form $(x-1)^n$ is independent.

(19) $af + bg$ belongs to $\mathcal{W}$ if and only if $a + b = 1$ so $\mathcal{W}$ is not a subspace.

2.3 Applications to Systems

1. T, 2. F, 3. T, 4. T, 5. F.

(1) Let A_i be the rows of A. Then $X = 4A_1 - 4A_2 + A_3$. Y is not in the row space.

(2) Reduced form:
$$R = \begin{bmatrix} 1 & 0 & 0 & 1 \\ 0 & 1 & 0 & 5 \\ 0 & 0 & 1 & -2 \\ 0 & 0 & 0 & 0 \end{bmatrix}$$

The basis is formed from the first three rows of R. The A may be expressed in terms of these rows as
$$\begin{aligned} A_1 &= 2R_1 + R_2 + 3R_3 \\ A_2 &= R_1 + R_2 + 3R_3 \\ A_3 &= R_2 + 2R_3 \\ A_4 &= 3R_1 + 3R_2 + 8R_3 \end{aligned}$$

(3) It is not in the row space.

(4) The reduced for of A^t is as follows. It has rank 3 as does A.
$$\begin{bmatrix} 1 & 0 & 0 & 1 \\ 0 & 1 & 0 & 1 \\ 0 & 0 & 1 & 1 \\ 0 & 0 & 0 & 0 \end{bmatrix}$$

(5) The rank of A is the is the maximal number of linearly independent columns in A. The columns of A become the rows of A^t, so the rank of A is the maximal number of linearly independent rows in A^t. But this is also the dimension of the row space of A^t which is the rank of A^t. Thus, A and A^t have the same rank.

(6) Basis: $\{[1,-2,0,2,3]^t, [1,-6,1,4,7]^t, [1,2,1,2,1]^t\}$

(7) Basis $X_1 = [1, 0, \frac{1}{11}, \frac{2}{11}]^t$, $X_2 = [0, 1, \frac{3}{11}, \frac{6}{11}]^t$.

(8) Basis: $\{[2,3,1,2]^t, [5,2,1,2]^t\}$

(9) Basis: $\{[1, 0, -5/3]^t, [0, 1, 7/3]^t\}$

(10) (a) The rank is two–the last column is twice the first and the third column is the first plus twice the second. The first two are independent. (b) $4 - 2 = 2$ (c) No. The column space is two dimensional (d) No–the nullspace is two dimensional. (e) Any two independent rows will work (f) The nullspace is the solution set to $AX = \mathbf{0}$. Since the row space is two dimensional, every equation in the corresponding system will be a linear combination of any two independent equations. The given equations correspond to rows 1 and 8.

(11) (a) The first two rows are independent and the last two are linear combinations of them. (The third row is the sum of the first two and the fourth is twice the first plus the second.) Thus the rank is two. (b) Any pair of independent columns is a basis. (c) The dimension of the nullspace is $6-2 = 4$ so the given vectors do not span the nullspace. (d) $T = [1, 0, 0, 0, 0, 0]^t$ (e) Any vector of the form $X = T + c_1 X_1 + c_2 X_2$ will work, where T is as in (d) and the X_i are as in (c).

(12) Yes. The dimension of the nullspace is $5-2 = 3$ and the given vectors are independent.

(13) The rank of A is $n - d$ so the dimension of the nullspace of A^t is $m - (n - d) = m - n + d$.

(14) The row space is $M(1. n)$, the column space is $R^n = M(n, 1)$ and the nullspace is $\mathbf{0}$.

(15) The dimension of the row space is less than or equal to m since A has only m rows. Therefore the rank is less then or equal to m. The same argument applied to the columns proves that the rank is also less than or equal to n.

2.3. APPLICATIONS TO SYSTEMS

(16) By definition, the rows span the row space. They are also independent since the dimension of the row space is m.

(17) The rank is less than or equal to the number of columns.

(18) No. Suppose that the entries of A are all non-zero, but its reduced form R has a row of zeros. Then every element of the column space of R will have a row of zeros, showing that the column space of R is not equal to the column space of A.

Chapter 3

Transformations

3.1 Matrix Transformations

True-False Answers: 1. T, 2. T, 3. F, 4. F, 5. T (Assuming that a point is a line segment of length 0.) 6. T, 7. T, 8. T, 9. F, 10. F.

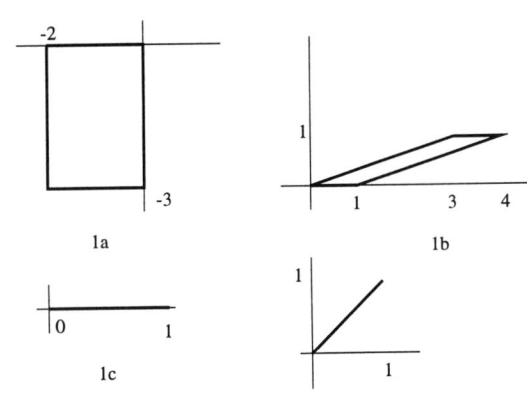

(1)

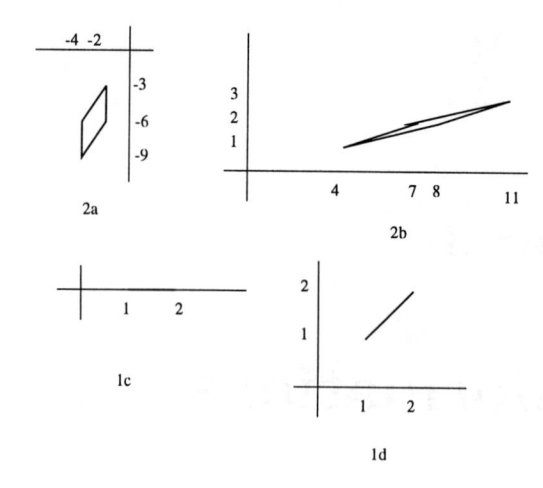

(2)

(3) $\frac{u^2}{4} + \frac{v^2}{9} = \frac{(2x)^2}{4} + \frac{(3y)^2}{9} = x^2 + y^2 = 1$

(4) It is the line $y = x$. The point $[x,y]^t$ is transformed into $[x,x]^t$ which is the point on the line lying either directly below or above the given point.

(5) Either $\begin{bmatrix} 2 & 4 \\ 4 & 2 \end{bmatrix}$ or $\begin{bmatrix} 4 & 2 \\ 2 & 4 \end{bmatrix}$.

(6)
$$\begin{bmatrix} -1/6 & 1/3 \\ 1/3 & -1/6 \end{bmatrix} \quad \text{and} \quad \begin{bmatrix} 1/3 & -1/6 \\ -1/6 & 1/3 \end{bmatrix}$$

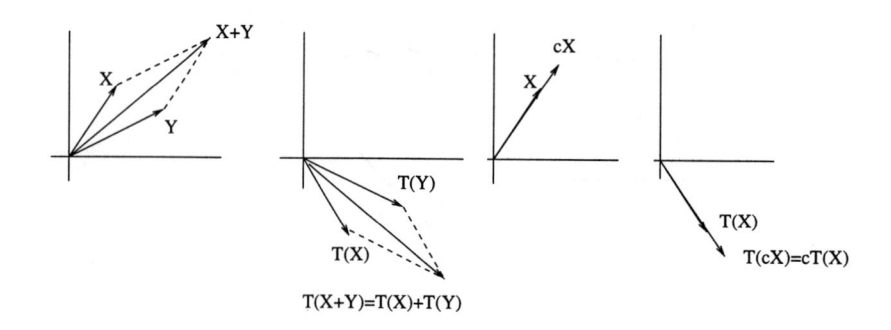

(7)

3.1. MATRIX TRANSFORMATIONS

(8) The figure is similar to that for Exercise 7.

(9) The figure is similar to that for Exercise 7.

(10) The transformation rotates θ radian about the z-axis.

(11) The transformation rotates θ radian about the x-axis.

(12) The transformation rotates θ radian about the y-axis.

(13) No, $X_2 = 2X_1$ but $Y_2 \neq 2Y_1$.

(14) No, $X_3 = X_1 + X_2$ but $Y_3 \neq Y_1 + Y_2$.

(15) (a) $[2,1,1]^t = 2[1,1,1]^t - [0,1,1]^t$ so $[2,1,1]^t$ is transformed onto $2[1,-2]^t - [1,1]^t = [1,-5]^t$ (b) $[2,2,3]^t = 2[1,1,1]^t + [0,0,1]^t$ so $[2,2,3]^t$ is transformed onto $2[1,-2]^t + [3,-5]^t = [5,-9]^t$ (c) $A = \begin{bmatrix} 0 & -2 & 3 \\ 3 & 0 & -5 \end{bmatrix}$.

(16) Since $[3,-5]^t = 19/8\,[2,-1]^t - 7/8\,[2,3]^t$, multiplication by A transforms $[3,-5]^t$ to $19/8\,Y_1 - 7/8\,Y_2$.

(17)

(a) $\begin{bmatrix} 1 & 0 & 0 & 0 \\ 0 & 1 & 2 & 0 \\ 0 & 0 & 1 & 0 \\ 0 & 0 & 0 & 1 \end{bmatrix}$ (b) $\begin{bmatrix} 1 & 0 & 0 \\ 0 & 17 & 0 \\ 0 & 0 & 1 \end{bmatrix}$ (c) $\begin{bmatrix} 0 & 1 & 0 & 0 \\ 1 & 0 & 0 & 0 \\ 0 & 0 & 1 & 0 \\ 0 & 0 & 0 & 1 \end{bmatrix}$

(18) (a) Any pair of points will work as long as neither entry is 0. For (b), see the following figure.

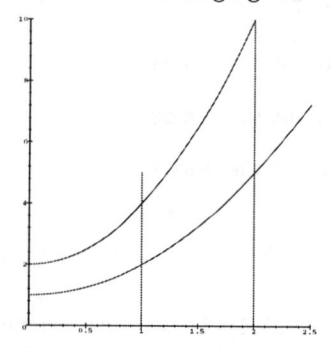

(19) (a)

$$(a) \begin{bmatrix} 2 & 3 & -7 \\ 0 & 0 & 0 \end{bmatrix} \quad (b) \begin{bmatrix} 1 & 1 \\ 1 & -1 \\ 1 & 2 \end{bmatrix}$$

$$(c) \begin{bmatrix} 1 & 0 & 0 & 0 & 0 \\ 0 & 1 & 0 & 0 & 0 \\ 0 & 0 & 1 & 0 & 0 \end{bmatrix}$$

(20) The image is the x-axis. The plane maps onto **0**. The set in question is a plane parallel to the stated plane.

(21) $R_{(-\theta)}$.

(23) (a) $T(f) = e^1 - e^{-1}$, $T(g) = 2/3$, $2T(f) + 3T(g) = 2e^1 - 2e^{-1} + 2$. (b) T is linear. (c) $2S(e^x) = e^2 - e^{-2}$, $S(2e^x) = 2(e^2 - e^{-2})$. (d) U is linear.

(24) (a) $D(2f + 3g) = 2e^x + 6x$. (b) D is linear (c) $S(e^x + x^2) = e^{2x} + 2e^x x^2 + x^4$ $S(e^x) + S(x^2) = e^{2x} + x^4$ (d) U is linear.

(25) Remark: The students might find it easier to think in terms of the "one is a linear combination of the others" definition of independence, rather than the "linear combinations equal to zero" definition.

3.2 Matrix Multiplication

True-False Answers: 1. F, 2. F, 3. T, 4. T, 5. F, 6. T, 7. T, 8. F.

3.2. MATRIX MULTIPLICATION

(1) $BC = \begin{bmatrix} 8 & 13 & -3 \\ -2 & -3 & 1 \end{bmatrix}$, $AB = \begin{bmatrix} 2 & 1 \\ 4 & 4 \end{bmatrix}$, $A(BC) = (AB)C = \begin{bmatrix} 4 & 7 & -1 \\ 12 & 20 & -4 \end{bmatrix}$

(2) Only $B(AC)$ is defined.

(3) $(AB)^t = \begin{bmatrix} 2 & 4 \\ 1 & 4 \end{bmatrix} = B^t A^t$. $A^t B^t = \begin{bmatrix} 8 & -2 \\ 10 & -2 \end{bmatrix}$.

(4) $B^t C^t$ is not defined.
$$C^t B^t = \begin{bmatrix} 8 & -2 \\ 13 & -3 \\ -3 & 1 \end{bmatrix}$$

(5) $(ABC)^t = C^t(AB)^t = C^t(B^t A^t) = C^t B^t A^t$.

(6) $(A_1 A_2 \ldots A_n)^t = A_n^t \ldots A_2^t A_1^t$.

(8)
$$\begin{aligned} A(C+D) &= [A(C_1+D_1), A(C_2+D_2), \ldots, A(C_q+D_q)] \\ &= [AC_1+AD_1, AC_2+AD_2, \ldots, AC_q+AD_q] \\ &= [AC_1, AC_2, \ldots, AC_q] + [AD_1, AD_2, \ldots, AD_q] \\ &= AC + AD \end{aligned}$$

(9) (a) Let $C = [c_1, \ldots, c_n]^t$. Then

$$\begin{aligned} (A+B)C &= c_1(A_1+B_1) + \ldots + c_n(A_n+B_n) \\ &= (c_1 A_1 + \ldots + c_n A_n) + (c_1 B_1 + \ldots + c_n B_n) = AC + BC \end{aligned}$$

. (b)

$$\begin{aligned} (A+B)C &= (A+B)[C_1, \ldots, C_q] \\ &= [(A+B)C_1, \ldots, (A+B)C_q] \\ &= [AC_1, \ldots, AC_q] + [BC_1, \ldots, BC_q] = AC + BC \end{aligned}$$

(10) (a) The ellipse $\frac{u^2}{4} + \frac{v^2}{9} = 1$. (b) The rectangle with vertices $[0,0]^t$, $[\sqrt{3}, 3/2]^t$, $[-1, \frac{3}{2}\sqrt{3}]^t$, and $[\sqrt{3}-1, 3/2+\frac{3}{2}\sqrt{3}]^t$ (c)
$$\begin{bmatrix} \sqrt{3} & -1 \\ 3/2 & 3/2\sqrt{3} \end{bmatrix}$$

(11) (a) The quadrilateral with vertices $[0,0]^t$, $[\sqrt{2}, \frac{\sqrt{2}}{2}]^t$, $[\sqrt{2}, \sqrt{2}]^t$, $[0, \frac{\sqrt{2}}{2}]^t$.

(b) $\begin{bmatrix} \sqrt{2} & 0 \\ \frac{\sqrt{2}}{2} & \frac{\sqrt{2}}{2} \end{bmatrix}$

(12) (a) The quadrilateral with vertices $[0,0]^t$, $[\sqrt{2}/, \sqrt{2}/2]^t$, $[\sqrt{3}, 3/2]^t$, $[\sqrt{2}/2, 3\sqrt{2}/2]^t$. (b) $\begin{bmatrix} \sqrt{2}/2 & 0 \\ \sqrt{2}/2 & \sqrt{2} \end{bmatrix}$

(13) (a) The ellipse $\frac{u^2}{16} + \frac{v^2}{81} = 1$. (b) $C = \begin{bmatrix} 4 & 0 \\ 0 & 9 \end{bmatrix}$.

(14) The image of the unit square under A^n is the parallelogram with vertices $[0,0]^t$, $[1,0]^t$, $[n,1]^t$ and $[n+1,1]^t$.

$$A^n = \begin{bmatrix} 1 & n \\ 0 & 1 \end{bmatrix}$$

(15) (a)

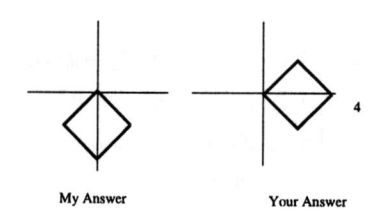

Exercise 15

(b) $M = \begin{bmatrix} \frac{\sqrt{2}}{2} & -\frac{\sqrt{2}}{2} \\ -\frac{\sqrt{2}}{2} & -\frac{\sqrt{2}}{2} \end{bmatrix}$, $Y = \begin{bmatrix} \frac{\sqrt{2}}{2} & \frac{\sqrt{2}}{2} \\ \frac{\sqrt{2}}{2} & -\frac{\sqrt{2}}{2} \end{bmatrix}$

(16) Since the nullspace is 1 dimensional, the columns of B are all multiples of a single column. For example:

$$B = \begin{bmatrix} -1 & -2 \\ 0 & 0 \\ 1 & 2 \end{bmatrix}$$

(18) All matrices of the form

$$\begin{bmatrix} a & b \\ 0 & a+b \end{bmatrix}$$

3.3. IMAGE OF A TRANSFORMATION

(19) On can let one of the matrices be the identity matrix, or let one be a multiple of the other, for example.

(20) Almost any randomly chosen A and B will work. The equality is true if and only if $AB = BA$.

(21) Almost any randomly chosen A and B will work. The equality is true if and only if $AB = BA$.

(22)
$$\begin{bmatrix} 0 & a \\ 0 & 0 \end{bmatrix} \quad \begin{bmatrix} 0 & 0 \\ a & 0 \end{bmatrix}$$

(23)
$$\begin{bmatrix} 0 & a & b \\ 0 & 0 & c \\ 0 & 0 & 0 \end{bmatrix}, \quad \begin{bmatrix} 0 & 0 & 0 \\ a & 0 & 0 \\ b & c & 0 \end{bmatrix}$$

(24) Any upper (or lower) triangular matrix with zeros on the main diagonal will work.

(25)
$$\begin{bmatrix} \pm 1 & 0 & 0 \\ 0 & \pm 1 & 0 \\ 0 & 0 & \pm 1 \end{bmatrix}$$

3.3 Image of a Transformation

True-False Answers: 1. T, 2. F, 3. T, 4. T, 5. F, 6. T, 7. F.

(1) (a) Rank: 3, image: $\{[1,2,2]^t, [2,3,10]^t, [0,2,4]^t\}$, nullspace: $\{[2,0,-7,1]^t\}$
(b) Rank: 3, image: $\{[-1,4,3]^t, [4,4,0]^t, [-2,2,3]^t\}$, nullspace: no basis (the nullspace is $\{\mathbf{0}\}$) (c) Rank: 2, , image: $\{[2,1,5]^t, [1,3,0]^t\}$, nullspace: $\{[-1,-3,5]^t\}$ (d) Rank: 1, image: $\{[1,2,3,-2]^t\}$, nullspace: $\{[-2,1,0]^t, [-2,0,1]^t\}$

(2) Since the rank is 3, the image is all of R^3.

(3) No. The pivot columns of the reduced form of A are $B_1 = [1,0,0]^t$ and $B_2 = [0,1,0]^t$. The system corresponding to $AX = B_1$ has reduced form as follows. Thus, the system is inconsistent, showing that B_1 is not in the image. Similarly, it follows that B_2 is not in the image.

$$\begin{bmatrix} 1 & 0 & 1/5 & 0 \\ 0 & 1 & 3/5 & 0 \\ 0 & 0 & 0 & 1 \end{bmatrix}$$

(4) From the solution to Exercise 11 in Section 2.3, this matrix has rank 2. Hence, any independent pair of columns (such as $[1,1,2,3]^t$ and $[7,1,8,15]^t$) will form a basis for the column space. Since the image is only two dimensional, the equation $AX = B$ is not solvable for all B. The dimension of the nullspace is $6 - 2 = 4$ so the solution is not unique.

(5) Make each column a multiple of $[1,2,3]^t$. Since the image is only 1 dimensional, the equation $AX = B$ is not solvable for all B. The dimension of the nullspace is $4 - 1 = 3$ so the solution is not unique.

(6) Let $X = [1,0,1]^t$ and $Y = [3,-1,0]^t$. Each column A_i of A must be of the form $a_i X + b_i Y$ where a_i and b_i are non-zero scalars. Furthermore, two of the columns must be independent. The rank A is two so the dimension of the nullspace is $5 - 2 = 3$.

(7) Each column of A must be a multiple of $[1,2]^t$ with at least one column non-zero. The dimension of the nullspace is $3 - 1 = 2$.

(8) Let the first two columns of A be non-zero multiples of $[1/2, 0, 1]^t$ and $[-1/2, 1, 0]^t$ and let the other columns be linear combinations of these vectors.

(9) No, the image must contain the zero vector.

Size	Always Solvable	Unique Solution	Dim. of Image	Dim. of Nullsp.	Rank
4x3	yes		imposible		
3x4	yes	no	3	1	3
5x5	no	no	4	1	4
5x5	yes	yes	5	0	5
3x2	no	yes	2	0	2
4x4	yes	yes	4	0	4
5x4	no	no	3	1	3
5x4			imposible		5

Figure 5

(10)

(11) The fourth and sixth rows.

(12) Since A and A^t have the same rank, one will be invertible if and only if the other is.

3.3. IMAGE OF A TRANSFORMATION

(13) The rank of A is n, the dimension of the image is n, and the nullspace of A has dimension $m - n$. Thus, $m \geq n$.

(14) (a) Let the columns of B be B_i and those of C be C_i. Then, $B_1 = -2B_2 + B_3$. Hence

$$C_1 = AB_1 = -2AB_2 + AB_3 = -2C_2 + C_3$$

(b) Let B have rank r and let $B_{i_1}, B_{i_2}, ..., B_{i_r}$ be a set of columns of B which forms a basis for the column space of B. Then, if B_i is any column, we may write $B_i = k_1 B_{i_1} + ... + k_r B_{i_r}$. Then

$$\begin{aligned} C_i &= AB_i = k_1 AB_{i_1} + ... + k_r AB_{i_r} \\ &= k_1 C_{i_1} + ... + k_r C_{i_r} \end{aligned}$$

Thus, the dimension of the column space of C is at most r.

(c) Now, let the rows of A be A_i and those of C be C_i. Then, $A_3 = 3A_2 - 6A_1$. Hence

$$C_3 = A_3 B = 3A_2 B - 6A_1 B = 3C_2 - 6C_1$$

(d) Part (d) is similar to (b).

(15) Let k and l be the, respectively, the dimension of the nullspace of B and AB. Then, the respective ranks of these matrices are $n - k$ and $n - l$ where n is the number of columns in B (and therefore in AB). The nullspace of B is contained in that of AB. Hence $k \leq l$ so $n - k \geq n - l$, as desired.

(16) $ABX = 0$ if and only if BX belongs to the nullspace of A which is equivalent with $BX = 0$ which is equivalent with X being in the nullspace of B. Thus, B and AB have the same nullspace. The rank is the number of columns minus the dimension of the nullspace so AB and B have the same rank.

(20) It suffices to show that the Y_i from Exercise 19 are independent which is a consequence of the fact that the nullspace of A is trivial.

3.4 Inverses

True-False Answers: 1. T, 2. T, 3. T, 4. T, 5. F, 6. F.

(1) $\begin{bmatrix} 2 & -1 & 0 \\ 0 & 2 & -1 \\ -1 & -1 & 1 \end{bmatrix}$

(2) (c) and (e) are not invertible. The inverses of the others are:

(a) $\begin{bmatrix} -\frac{13}{5} & 3/2 & -6/5 \\ 2 & -1 & 1 \\ 6/5 & -1/2 & 2/5 \end{bmatrix}$ (b) $\begin{bmatrix} -1/2 & 1/2 & 1/2 \\ 1/2 & -1/2 & 1/2 \\ 1/2 & 1/2 & -1/2 \end{bmatrix}$

(d) $\begin{bmatrix} 1/2 & -1/4 & 1/4 \\ -1/2 & 5/4 & -1/4 \\ 0 & -1/2 & 1/2 \end{bmatrix}$ (f) $\begin{bmatrix} \frac{23}{3} & \frac{17}{3} & -\frac{17}{3} & -13/3 \\ 2/3 & 2/3 & -2/3 & -1/3 \\ 3 & 3 & -2 & -2 \\ -5 & -4 & 4 & 3 \end{bmatrix}$

(g) $\begin{bmatrix} 1 & -1/2 & 0 & 0 \\ 0 & 1/2 & -1/3 & 0 \\ 0 & 0 & 1/3 & -1/4 \\ 0 & 0 & 0 & 1/4 \end{bmatrix}$ (h) $\begin{bmatrix} a^{-1} & 0 & 0 & 0 \\ 0 & b^{-1} & 0 & 0 \\ 0 & 0 & c^{-1} & 0 \\ 0 & 0 & 0 & d^{-1} \end{bmatrix}$

(3) (a) $[-\frac{16}{5}, 3, -\frac{7}{5}]^t$, (b) $[2, 1, 0]^t$, (c) not invertible, (d) $[\frac{3}{4}, \frac{5}{4}, \frac{1}{2}]^t$, (e) not invertible (f) $[-\frac{46}{3}, -\frac{4}{3}, -5, 11]^t$, (g) $[0, 0, 0, 1]^t$, (h) $[\frac{1}{a}, \frac{2}{b}, \frac{3}{c}, \frac{4}{d}]^t$

(4) (c) The second row is twice the first and the third row is three times the first. The columns follow the same pattern. (e) Let A_i be the columns and B_i the rows. Then $B_4 = B_1 + B_3$ and $A_4 = -817/13\, A_1 + 149/13\, A_2 + 355/13\, A_3$.

(5) (c) and (e) are not invertible.

(6)
$$X = \begin{bmatrix} \frac{50}{9} & 5/9 & -\frac{17}{9} \\ -\frac{11}{9} & -2/9 & 5/9 \\ -\frac{8}{9} & 1/9 & 2/9 \end{bmatrix} \begin{bmatrix} 1 \\ 2 \\ 7 \end{bmatrix} = \begin{bmatrix} -\frac{59}{9} \\ \frac{20}{9} \\ \frac{8}{9} \end{bmatrix}$$

3.4. INVERSES

(7) It fails because the coefficient matrix is not invertible. In fact the third row is twice the first plus the second.

(8) Make one of the rows (or columns) be a linear combination of the others.

(9) $\begin{bmatrix} 1 & -a & ac-b \\ 0 & 1 & -c \\ 0 & 0 & 1 \end{bmatrix}$

(10) Let A an $n \times n$, upper triangular, unipotent matrix and let $B = [A, I]$ be the double matrix which we reduce to compute A^{-1}. Since the diagonal entries of A are already 1, we can reduce A to I by subtracting multiples of each row of B from rows lying above it. This will not change any entries of the right side of B which lie on or below the diagonal. Hence, A^{-1} will have all of its diagonal entries equal to 1 and all of the entries below the diagonal equal to 0, proving that A^{-1} is unipotent.

(11) $\frac{1}{ad-bc} \begin{bmatrix} d & -b \\ -c & a \end{bmatrix}$

(12)
$$X = CB = \begin{bmatrix} 16 & 9 & 8 \\ 6 & 3 & 4 \\ 0 & 0 & 0 \end{bmatrix}$$

(13)
$$X = BC = \begin{bmatrix} 3 & 4 & 7 \\ 2 & 2 & 4 \\ 6 & 8 & 14 \end{bmatrix}$$

(14) This follows directly from Theorem 2 and the fact that $A^{-1}A = I$.

(15) If $ABX = Y$ then $BX = A^{-1}Y$ so $X = B^{-1}(A^{-1}Y) = (B^{-1}A^{-1})Y$. Hence $(AB)^{-1} = B^{-1}A^{-1}$.

(16)
$$\begin{aligned}(B^{-1}A^{-1})(AB) &= B^{-1}A^{-1}AB \\ &= B^{-1}IB = B^{-1}B = I\end{aligned}$$

(17) Similar to 16.

(18) The inverses are, respectively, $(A^{-1})^2$, $(A^{-1})^3$, and $(A^{-1})^4$.

(19) $(A^{-1})^t A^t = (AA^{-1})^t = I^t = I$.

(20) Multiply $ABAB = AABB$ on the left by A^{-1} and on the right by B^{-1}.

(22) $A^n = QD^nQ^{-1}$ which we compute as:

$$Q^{-1} = \begin{bmatrix} 1/2 & -1/2 \\ 1/2 & 1/2 \end{bmatrix} \quad A^n = \begin{bmatrix} 2^{n-1} + (1/2)\,4^n & -2^{n-1} + (1/2)\,4^n \\ -2^{n-1} + (1/2)\,4^n & 2^{n-1} + (1/2)\,4^n \end{bmatrix}$$

(23) (a) $B = \begin{bmatrix} -1 & 2 \\ 1 & -1 \\ 0 & 0 \end{bmatrix}$ (Other answers are possible.) (b) Choose $B = [B_1, B_2]$ where $AB_1 = I_1$ and $AB_2 = I_2$. (c) $2 = \text{rank}(I) = \text{rank}(AB) \leq \text{rank}(A) \leq 2$. (d) An $m \times n$ matrix has a left inverse if and only if its rank is m.

(24) Let $B' = B^t$ where B is as in Exercise 23a. An $m \times n$ matrix has a left inverse if and only if it has rank n.

(25) Let B be the left inverse of A. If $AX = 0$, then $0 = BAX = X$, showing that the nullspace is zero. A proof based on the rank of products theorem is also possible.

(26) See the answer to Exercise 14, Section 3.3.

(27) See the answer to Exercise 14, Section 3.3.

(29) $I = -A^2 - 3A = A(-A - 3I)$. Thus, from Theorem 2, $A^{-1} = -A - 3I$.

(30) $I = A(-A^2 - 3A - 2I)/5$. Thus, from Theorem 2, $A^{-1} = (-A^2 - 3A - 2I)/5$.

(31) (b) $(I - N)(I + N + N^2) = I - N^4 = I$. Thus, from Theorem 2, $(I - N)^{-1} = I + N + N^2$. (c) is similar.

3.5 The LU Factorization

(1) (a) $[-8, 3, -7, 6]^t$ (b) $[5, -7 - x_4, 3 - x_4, x_4]^t$

3.5. THE LU FACTORIZATION

(2)
$$A = \begin{bmatrix} 1 & 0 & 0 \\ 2 & 1 & 0 \\ 1 & 1/2 & 1 \end{bmatrix} \begin{bmatrix} 1 & 4 & -1 \\ 0 & 2 & 4 \\ 0 & 0 & 1 \end{bmatrix}$$

$$B = \begin{bmatrix} 1 & 0 & 0 & 0 \\ 2 & 1 & 0 & 0 \\ 1 & 2 & 1 & 0 \\ 3 & -1 & -\frac{7}{8} & 1 \end{bmatrix} \begin{bmatrix} 1 & 2 & 0 & 1 \\ 0 & -1 & 4 & -1 \\ 0 & 0 & -8 & 5 \\ 0 & 0 & 0 & \frac{35}{8} \end{bmatrix}$$

(3) You must interchange rows 2 and 3 of A for it to work. The **LU** decomposition for the resulting matrix is

$$\begin{bmatrix} 1 & 0 & 0 \\ 2 & 1 & 0 \\ 2 & 0 & 1 \end{bmatrix} \begin{bmatrix} 1 & 1 & 3 \\ 0 & -1 & -2 \\ 0 & 0 & -1 \end{bmatrix}$$

(4) In the **LU** process, we only subtract rows from lower rows. Thus, the 0 in the $(2,2)$ position of A resulted from subtracting a multiple of the first row from the second. If we had exchanged the second and third rows before beginning, we would have obtained a 1 in this position which would allow us to proceed. We might still need to exchange the new second and third rows to complete the process.

(5)
$$A = \begin{bmatrix} 1 & 0 & 0 \\ 2 & 1 & 0 \\ 1 & -1/3 & 1 \end{bmatrix} \begin{bmatrix} 1 & 0 & 0 \\ 0 & -3 & 0 \\ 0 & 0 & 2/3 \end{bmatrix} \begin{bmatrix} 1 & 2 & 1 \\ 0 & 1 & 1/3 \\ 0 & 0 & 1 \end{bmatrix}$$

(6)
$$\begin{bmatrix} 1 & 0 & 0 \\ -a & 1 & 0 \\ ac-b & -c & 1 \end{bmatrix}$$

(7) See the answer to Exercise 10 in Section 3.4. Only a few words need to be changed since we are now discussing lower triangular matrices.

(8) Multiplication of a vector Y by B applies all of the steps used in reducing A to Y. Thus, multiplication of A by B will apply these same steps to each column of A, which is how U was produced in the first place.

(9) If $[U, B]$ is produced by reducing $[A, I]$, then $BA = U$.

Chapter 4

Orthogonality

4.1 Coordinates

True-False Answers: 1. F, 2. F, 3. F, 4. F, 5. T.

(1) (a) Let $X = [1,1]^t$ and $Y = [1,-1]^t$. Then the equation is $1 = (x')^2 - (y')^2$, which is of the desired form. (b) With X and Y as in (a), the equation is $1 = 3(x')^2 + (y')^2$ (c) $25(x')^2 + 75(y')^2 = 1$. The curve is an ellipse with intercepts $x = \pm 1/5$ and $y = \pm 1/(5\sqrt{5})$. The width is $2/\sqrt{5}$. (The width of the original figure is $(2/\sqrt{5})|[1,2]| = 2$.)

(2) Below, P the point matrix, C is the coordinate matrix, and X is the coordinate vector for $[1,2,3]^t$

(a) $P = \begin{bmatrix} 1 & 0 & 0 \\ 1 & 1 & 0 \\ 1 & 1 & 1 \end{bmatrix}$ $C = \begin{bmatrix} 1 & 0 & 0 \\ -1 & 1 & 0 \\ 0 & -1 & 1 \end{bmatrix}$ $X = \begin{bmatrix} 1 \\ 1 \\ 1 \end{bmatrix}$

(b) $P = \begin{bmatrix} 1 & 2 & 1 \\ -2 & 3 & 1 \\ 1 & 2 & 0 \end{bmatrix}$ $C = \begin{bmatrix} 2/7 & -2/7 & 1/7 \\ -1/7 & 1/7 & 3/7 \\ 1 & 0 & -1 \end{bmatrix}$ $X = \begin{bmatrix} 1/7 \\ \frac{10}{7} \\ -2 \end{bmatrix}$

(c) $P = \begin{bmatrix} 1 & -1 & 5 \\ 3 & 1 & 1 \\ 2 & -1 & -4 \end{bmatrix}$ $C = \begin{bmatrix} 1/14 & 3/14 & 1/7 \\ -1/3 & 1/3 & -1/3 \\ \frac{5}{42} & 1/42 & -2/21 \end{bmatrix}$ $X = \begin{bmatrix} \frac{13}{14} \\ -2/3 \\ -\frac{5}{42} \end{bmatrix}$

CHAPTER 4. ORTHOGONALITY

(3) (a) The third basis (b) $X' = [\frac{13}{14}, -\frac{2}{3}, -\frac{5}{42}]^t$ (c) Same as (b) (d)

$$\begin{bmatrix} \frac{1}{14} & \frac{3}{14} & \frac{1}{7} \\ -\frac{1}{3} & \frac{1}{3} & -\frac{1}{3} \\ \frac{5}{42} & \frac{1}{42} & -\frac{2}{21} \end{bmatrix}$$

(4) The new coordinates are $X' = QX$ where Q is as follows. P is the point matrix.

$$P = \begin{bmatrix} \sqrt{2}/2 & \sqrt{3}/3 & \sqrt{6}/6 \\ \sqrt{2}/2 & -\sqrt{3}/3 & -\sqrt{6}/6 \\ 0 & \sqrt{3}/3 & -\sqrt{6}/3 \end{bmatrix} \quad Q = \begin{bmatrix} \sqrt{2}/2 & \sqrt{2}/2 & 0 \\ \sqrt{3}/3 & -\sqrt{3}/3 & \sqrt{3}/3 \\ \sqrt{6}/6 & -\sqrt{6}/6 & -\sqrt{6}/3 \end{bmatrix}$$

(5) $X' = [\frac{4}{7}, \frac{16}{35}, 0, -\frac{3}{5}]^t$

(6) $Q_4 = a[-1, 0, 1, 0]^t$ where a is ant non-zero scalar.

(7) Let M be the 3×4 matrix which has the Q_i as its rows. Since the Q_i are independent, M has rank three. Thus, the nullspace 1 dimensional. We may choose Q_5 to be any element of this nullspace. If the Q_i were in R^5, then M would have a 2-dimensional nullspace and Q_5 could be any non-zero element of this two dimensional space.

(10) Call the polynomials p_i. (a) Oops! The set is independent! (b) $p_3 = -p_1 + 2p_2$ (c) $p_5 = p_1 + 3p_2 - 4p_3$, $p_4 = 2p_1 - p_2 + 3p_3$.

(11) Let $Z' = aX' + bY'$. Then $z_i' = ax_i' + by_i'$ where z_i', x_i' and y_i' are the entries of Z', X' and Y' respectively. Hence,

$$\begin{align} Z &= z_1'Q_1 + \ldots + z_n'Q_n \\ &= ax_1'Q_1 + \ldots + ax_n'Q_n + by_1'Q_1 + \ldots + by_n'Q_n = aX + bY \end{align}$$

(12) (a) $[1, 1, 0]^t$ (b) $[5, 5, 1]^t$, (c) $[4, 1, 2]^t$

(13) (a) $[-3, 1, 0]^t$ (b) $[1, -3, 1]^t$ (c) $[22, -15, 2]^t$

4.2 Projections: The Gram-Schmidt Process

True-False Answers: 1. T, 2. F, 3. F, 4. T.

4.3. FOURIER SERIES: SCALAR PRODUCT SPACES

(1) Projection: $[\frac{47}{42}, \frac{85}{42}, \frac{40}{21}]^t$

(2)
$$\begin{bmatrix} \frac{17}{42} & -\frac{5}{42} & \frac{10}{21} \\ -\frac{5}{42} & \frac{41}{42} & 2/21 \\ \frac{10}{21} & 2/21 & \frac{13}{21} \end{bmatrix}$$

(3) Projection: $\frac{1}{5}[8, 4, 4, -12]^t$

(4) Projection $[1, 0, 1, 0]^t$. Hence X is in the subspace.

(5) (a) $\{[0, 1, 1]^t, [1, 0, 0]^t\}$ (b) $\{[1, 2, 1, 1]^t, [-2, 1, 1, -1]^t, \frac{1}{7}[-6, -2, 0, 10]$

(6) (a) $[1, 5/2, 5/2]^t$
    (b) $[\frac{13}{7}, \frac{16}{7}, 1, \frac{11}{7}]^t$

(7) (c) $\frac{(x+2y+z-w)}{7}[1, 2, 1, -1]^t + \frac{(-y+z-w)}{3}[0, -1, 1, -1]^t$

(8) (a)
$$Q = \begin{bmatrix} 1 & 0 \\ 1 & 1 \\ 0 & 1 \\ 1 & -1 \end{bmatrix} \qquad N = \begin{bmatrix} 1 & 1 \\ 0 & 1 \end{bmatrix}$$

(b)
$$Q = \begin{bmatrix} 2 & 2/7 & \frac{15}{13} \\ 3 & 3/7 & -\frac{10}{13} \\ 1 & -\frac{13}{7} & 0 \end{bmatrix} \qquad N = \begin{bmatrix} 1 & 6/7 & -1/14 \\ 0 & 1 & -1/26 \\ 0 & 0 & 1 \end{bmatrix}$$

(9) (a) $[-3, 1, 0, 0]^t, [-1, -3, 30, 20]^t$ (b) $\frac{1}{65500}[26223, 78669, 6431, 6504]^t$

(10) The answer is the same as in 9a.

(11) Basis: $[1, 3, 1, -1]^t, [5, 15, -19, 31]^t$

4.3 Fourier Series: Scalar Product Spaces

True-False Answers: 1. T, 2. F, 3. T, 4. T.

(2) (a) $\sinh x = \frac{e^x - e^{-x}}{2}$

(b) $e^{-x} = \cosh x - \sinh x$
(c) $\cos(2x) = \cos^2 x - \sin^2 x$
(d) $\cos^2 x = \frac{\cos(2x)}{2} + \frac{1}{2}$
(e) $\sin x = \frac{\sqrt{2}}{2}(\sin(x + \frac{\pi}{4}) - \cos(x + \frac{\pi}{4}))$
(f) $(x+3)^2 = x^2 + 6x + 9$
(g) $x^2 + 3x + 3 = 3(x+1) + \frac{1}{2}(2x^2)$

(5) $f_o(x) = \sum_{k=1}^n [\frac{12}{k^3\pi^3} - \frac{2}{k\pi}](-1)^k \sin k\pi x$

(6) All of the coefficients are 0.

(7) (a) For $n \neq \pm m$
$$\int_{-1}^{1} \cos(n\pi x)\cos(m\pi x)dx = 1/2 \left(\frac{\sin((n\pi - m\pi)x)}{n\pi - m\pi} + \frac{\sin((n\pi + m\pi)x)}{n\pi + m\pi} \right) \Big|_{-1}^{1} = 0$$

(b) For $n \neq 0$, (p_n, p_n) equals
$$\int_{-1}^{1} (\cos(n\pi x))^2 dx = \frac{\frac{1}{2}\cos(n\pi x)\sin(n\pi x) + \frac{1}{2}n\pi x}{n\pi} \Big|_{-1}^{1} = 1$$

Also (p_0, p_0) equals
$$\int_{-1}^{1} 1 \, dx = 2$$

(c) $f_o(x) = \frac{1}{3} + \sum_{k=1}^n [\frac{4}{k^2\pi^2}](-1)^k \cos k\pi x$

(8)
$$f_o(x) = \frac{1}{2} + \sum_{k=1}^{n} \frac{2((-1)^k - 1)}{k^2\pi^2} \cos k\pi x$$

(9) (a)
$$\int_{-1}^{1} \cos(n\pi x)\sin(m\pi x)dx = \left(-\frac{1}{2}\frac{\cos((n\pi + m\pi)x)}{n\pi + m\pi} + \frac{1}{2}\frac{\cos((n\pi - m\pi)x)}{n\pi - m\pi} \right) \Big|_{-1}^{1} = 0$$

(b)
$$f_o(x) = \frac{1}{4} + \sum_{k=1}^{n} \frac{((-1)^k - 1)}{k^2\pi^2} \cos k\pi x$$
$$+ \sum_{k=1}^{n} \frac{(-1)^{k+1}}{k\pi} \sin k\pi x$$

(10) $(1, x^2) = \frac{2}{3}$. Orthogonal set: $\{1, x, x^2 - 1/3\}$. Projection: $\frac{3}{5}x$.

(11) $c_k = \frac{(-1)^{k+1}(.002)}{k\pi}$

(13) (a) 8 (d) $c_1 = 0$, $c_2 = -1$, $c_3 = 2$.

(14) The set of all polynomials is a subspace but the set of polynomials with integral coefficients is not.

(17) $n + 1$

(18) Infinite.

4.4 Orthogonal Matrices

True-False Answers: 1. F, 2. T, 3. T, 4. T, 5. F, 6. F.

(1) $|AX| = |[7, \sqrt{2}, 3]^t| = \sqrt{60} = |X|$

(3) Compute $(R_\theta^x)^t R_\theta^x$, $(R_\theta^y)^t R_\theta^y$, and $(R_\theta^z)^t R_\theta^z$ to see that you do get I.

(4) $c = \frac{1}{7}$

(6) $|(AB)X| = |A(BX)| = |BX| = |X|$.

(7) $(AB)(AB)^t = ABB^t A^t = AA^t = I$.

(8) $A^{-1}(A^{-1})^t = A^t(A^t)^t = A^t A = A^{-1} A = I$.

(9) This exercise is equivalent with showing that A^t is orthogonal. But $A^t(A^t)^t = A^{-1}A = I$.

(10) The last column of A is $\pm[0, -\sqrt{2}/2, 0, \sqrt{2}/2]$

(12) No it is not possible. Since orthonormal sets are independent, the matrix would have rank 3 which is impossible.

(13) $A^t A = I$.

4.5 Least Squares

True-False Answers: 1. T, 2. F, 3. T, 4. F, 5. F.

(1) $X = [30.9306]^t$, $B_0 = [31.9537, 33.1815, 34.0, 35.2278, 35.637]^t$

(2) The system is $AX = B$ where $X = [a, b]^t$ and

$$A = \begin{bmatrix} 0.5000 & -0.6931 \\ 1.1000 & 0.0953 \\ 1.5000 & 0.4055 \\ 2.1000 & 0.7419 \\ 2.3000 & 0.8329 \end{bmatrix} \quad B = \begin{bmatrix} 32.0000 \\ 33.0000 \\ 34.2000 \\ 35.1000 \\ 35.7000 \end{bmatrix}$$

Then

$$A^t A = \begin{bmatrix} 13.4100 & 3.8402 \\ 3.8402 & 1.8981 \end{bmatrix} \quad A^t B = \begin{bmatrix} 259.4200 \\ 50.6083 \end{bmatrix}$$

yielding $X = [27.8393, -29.6609]^t$ and
$AX = [34.4790, 27.7962, 29.7324, 36.4559, 39.3255]^t$. Also $|AX - M| = 8.2563$. In Exercise 1, $|AX - M| = 0.3088$ so the linear model produces the closer approximation.

(3) (a)

$$A = \begin{bmatrix} 1 & .5 & .25 \\ 1 & 1.1 & 1.21 \\ 1 & 1.5 & 2.25 \\ 1 & 2.1 & 4.41 \\ 1 & 2.3 & 5.29 \end{bmatrix} \quad B = \begin{bmatrix} 32.0, 33.0, 34.2, 35.1, 35.7 \end{bmatrix}$$

(b) $[a, b, c]^t = [30.9622, 1.98957, .019944]^t$.

$$A^t A = \begin{bmatrix} 5.0000 & 7.5000 & 13.4100 \\ 7.5000 & 13.4100 & 26.2590 \\ 13.4100 & 26.2590 & 54.0213 \end{bmatrix} \quad A^t B = \begin{bmatrix} 170.0000 \\ 259.4200 \\ 468.5240 \end{bmatrix}$$

(4) The system is $AX = B$ where $X = [a, b]^t$,

$$A = \begin{bmatrix} 0 & 1 \\ 5 & 1 \\ 10 & 1 \\ 15 & 1 \end{bmatrix} \quad B = \begin{bmatrix} 0 \\ 10 \\ 22 \\ 35 \end{bmatrix}$$

4.5. LEAST SQUARES

Then
$$A^t A = \begin{bmatrix} 350 & 30 \\ 30 & 4 \end{bmatrix} \qquad A^t B = \begin{bmatrix} 795 \\ 67 \end{bmatrix}$$

Producing $[a,b]^t = [2.3400, -0.8000]^t$. The predicted population is $30a + b + 227 = 69.4000 + 227 = 296.4000$.

(5) The system is $AX = B$ where $X = [\ln a, b]^t$,

$$A = \begin{bmatrix} 1 & 0 \\ 1 & 5 \\ 1 & 10 \\ 1 & 15 \end{bmatrix} \qquad B = \begin{bmatrix} 5.4250 \\ 5.4681 \\ 5.5175 \\ 5.5683 \end{bmatrix}$$

Then
$$A^t A = \begin{bmatrix} 4 & 30 \\ 30 & 350 \end{bmatrix} \qquad A^t B = \begin{bmatrix} 21.9788 \\ 166.0400 \end{bmatrix}$$

Producing $[\ln a, b]^t = [5.4228, 0.0096]^t$. Then $P = 226.512 e^{.00956 t}$. In 2010, $P = 301.750$.

(6) The projection is $B_0 = [-2/7, 37/14, 15/14]^t$. The distance is $|B - B_0| = \frac{9}{14}\sqrt{14}$.

(7) $B_0 = \frac{1}{14}[3, 4, 2, 3, 2]^t$

(8)
$$P = \frac{1}{14} \begin{bmatrix} 3 & 4 & 2 & 3 & 2 \\ 4 & 10 & -2 & 4 & -2 \\ 2 & -2 & 6 & 2 & 6 \\ 3 & 4 & 2 & 3 & 2 \\ 2 & -2 & 6 & 2 & 6 \end{bmatrix}$$

(9) A is not invertible.

(10) Let $A = [A_1, A_2]$ and $C = [C_1, C_2]$ be the 2×5 matrices which have the indicated vectors as columns. Then ther is an invertible 2×2 matrix B such that $C = AB$. Then

$$C(C^t C)^{-1} C^t = AB(B^t A^t AB)^{-1} B^t A^t = A(A^t A)^{-1} A^t$$

(11) Since each A_i is a linear combination of the C_i, there is an invertible matrix B such that $A = CB$ where A has the A_i as columns and B has the B_i as columns. Thus, the argument from Exercise 10 applies.

(12) The rank is k and the nullspace is the orthogonal complement of $\mathcal{W}$.

(15) (a) In C, the third row is the sum of the first two (b) In the augmented matrix, the third row is the sum of the first two

$$C = \begin{bmatrix} 1 & 2 & 3 \\ 3 & 1 & 4 \\ -2 & 1 & -1 \end{bmatrix} \qquad A^t B = \begin{bmatrix} 1 \\ 7 \\ 8 \end{bmatrix}$$

(17) (a) $S^\perp$ is the line through the origin perpendicular to the original line, $(S^\perp)^\perp$ is the original line. (b) $S^\perp$ is the line through the origin which is perpendicular to the original vectors, $(S^\perp)^\perp$ is the plane spanned by the vectors. (c) $S^\perp$ is the line through the origin perpendicular to the original plane, $(S^\perp)^\perp$ is the original plane. (d) $S^\perp$ is the plane through the origin perpendicular to the original line, $(S^\perp)^\perp$ is the original line.

Chapter 5

Determinants

5.1 Determinants

1. T, 2. F, 3. F, 4. T, 5. T, 6. T.

(1) (a) 14 (b) 0 (c) 0 (d) 7(ac-db) (e) -16 (f) 0 (g) 0 (h) 16 (i) 34 (j) -6 (k) 144 (l) -2

(2) (b) $R_2 = 2R_1$ (c) $R_3 = 2R_2$ (f) $R_3 = R_1 + R_2$ (g) $R_3 = 2R_2 - R_1$

(3)

$$-a_{21}\left(a_{12}a_{33} - a_{13}a_{32}\right) + a_{22}\left(a_{11}a_{33} - a_{13}a_{31}\right) - a_{23}\left(a_{11}a_{32} - a_{12}a_{31}\right)$$

(4)

$$\alpha = \begin{vmatrix} -13 & 7 & 9 & 5 \\ 16 & -37 & 99 & 64 \\ -42 & 78 & 55 & -3 \\ 47 & 29 & -14 & -8 \end{vmatrix} - 4 \begin{vmatrix} 24 & 7 & 9 & 5 \\ 11 & -37 & 99 & 64 \\ 31 & 78 & 55 & -3 \\ 62 & 29 & -14 & -8 \end{vmatrix}$$

$$+ 2 \begin{vmatrix} 24 & -13 & 9 & 5 \\ 11 & 16 & 99 & 64 \\ 31 & -42 & 55 & -3 \\ 62 & 47 & -14 & -8 \end{vmatrix} - 2 \begin{vmatrix} 24 & -13 & 7 & 5 \\ 11 & 16 & -37 & 64 \\ 31 & -42 & 78 & -3 \\ 62 & 47 & 29 & -8 \end{vmatrix}$$

$$-3\begin{vmatrix} 24 & -13 & 7 & 9 \\ 11 & 16 & -37 & 99 \\ 31 & -42 & 78 & 55 \\ 62 & 47 & 29 & -14 \end{vmatrix}$$

The expansion of β is similar except that each coefficient is multiplied by 3.

(5) The 4×4 determinants are the same as in Exercise 4. Note: $\alpha = \gamma - \delta$, not $\gamma + \delta$ as stated in the exercise.

(10) 0, 2, 0, 0.

5.2 Reduction and Determinants

True-False Answers: 1. T, 2. T, 3. T, 4. T, 5. T, 6. F.

(1) (a) 0 (b) 16 (c) 34 (d) -6 (e) 0

(2) See the answers to Exercise 1.

(3) See the answers for Exercise 1, Section 5.1.

(4) -30

(5) 120 and -5

(9) The students should do Exercise 8 first.

(11) See the argument on p. 286.

5.3 A Formula for Inverses

True-False Answers: 1. T, 2. T.

(1) $x = -\frac{9}{4}$, $y = -\frac{5}{4}$, $z = 2$, $w = \frac{15}{4}$

(2) $z = -1/9\, p_1 - 1/9\, p_2 + 1/9\, p_3$

(3) $c_1(x) = e^{-x}(1 + x^2)\cos x - e^{-x} x \sin x$, $c_2(x) = x \cos x + (1 + x^2)\sin x$

(4) (b) 1, (c) 0, (d) 5/2

5.3. A FORMULA FOR INVERSES

(5) -1

(6)
$$A^{-1} = \frac{1}{2}\begin{bmatrix} 0 & -1 & 2 \\ 2 & 3 & -8 \\ -2 & -1 & 6 \end{bmatrix}$$

(7) You get A.

(8)
$$(-x^4 + 2x^9 - x^{14})^{-1}\begin{bmatrix} x^4 - x^2 & x^8 - x^5 & x^4 - x^9 \\ x^6 - x^3 & -x^{12} + x^4 & x^8 - x^3 \\ -x^7 + x^2 & x^8 - x^3 & 0 \end{bmatrix}$$

Chapter 6

Eigenvectors

6.1 Eigenvectors

True-False Answers: 1. T, 2. T, 3. F, 4. T, 5. F.

(1) (a) X is an eigenvector with eigenvalue 3, Y is not an eigenvector (b) both X and Y are eigenvectors with eigenvalues 0 and 3 respectively (c) X is an eigenvector with eigenvalue -1, Y is not. (By definition, eigenvectors must be non-zero.)

(2) (a) The eigenvalues for X, Y and Z are, respectively, 6, -3, and 0. Also, $B = Y + Z$ so $A^{10}B = (-3)^{10}Y = [59049, 59049]^t$.

(b) The eigenvalues for X, Y and Z are, respectively, 2, 2, and 4. Also, $B = (3/2)X - (1/2)Y - (1/2)Z$ so
$A^{10}B = 3(2)^9 X - 2^9 Y - 2(4^9)Z = [1572352, 1572352, 1573376]^t$.

(3) Any vector $aX + bY$ as long as $a > 0$ and $b > 2a$ will work.

(5) One eigenspace is the xy plane and the other is the z-axis.

(6) False. They must corespond to the same eigenvalue.

(7) The nullspace.

(8) False: if the nullspace is non-zero, the matrix canot be invertible.

(9) (a) $p(\lambda) = -\lambda^2(\lambda - 3)$, eigenvalues and corresponding basis: $\lambda = 0$: $[1, 0, 1]^t$ and $[1, -1, 0]^t$, $\lambda = 3$, $[1, 1, 1]^t$, $A^n B = 5[3^{n-1}, 3^{n-1}, 3^{n-1}]^t$ for $n > 1$.

(10) (a) 5, 8, 13 (b) $A^n X = [F_{n+1}, F_{n+2}]$ (c) The eigenvalues and eigenvectors are, respectively, $\lambda_\pm = \sqrt{5}/2 \pm 1/2$ and $X_\pm = [1, \lambda_\pm]^t$. We find $[1,1]^t = x_+ X_+ + x_- X_-$ where $x_\pm = 1/2 \pm \sqrt{5}/10 = \pm \lambda_\pm / \sqrt{5}$

(c) $A^{10} X = [89, 144]^t$ so $F_{11} = 89$ and $F_{12} = 144$. However, to compute F_{10} we should compute $A^8 X = [34, 55]^t$ showing that $F_{10} = 55$. (d)
$$A^n X = \lambda_+^{n+1} [1, \lambda_+]^t / \sqrt{5} - \lambda_-^{n+1} [1, \lambda_-]^t / \sqrt{5}$$
Thus, $F_n = (\lambda_+^n - \lambda_-^n)/\sqrt{5}$

(11) (a) $\lambda = -8$, basis: $[-1, -1, 5]^t$. This matrix is deficient over the real numbers (but not over the complex numbers.) (b) is deficient. The eigenvalues are 2 (multiplicity 2) and 1 with respective eigenvectors $[2, 1, 2]^t$ and $[1, 0, 1]^t$. (c) The eigenvalues are 4 (basis $[1, 1, 0, 0]$), -2 (basis $[1, -1, 0, 0]^t$ and $[0, 0, 2, -1]^t$) and -3 (basis $[0, 0, -1, 1]^t$.)

(14) The coefficient of λ^n is $(-1)^n$.

(15) (a) 10×10 (b) No. (c) 1, 2, or 3. (d) $1 \le d \le 5$.

(16) Any matrix with trace 5 and determinant 6 will work. E.g.
$$\begin{bmatrix} 2 & a \\ 0 & 3 \end{bmatrix}$$

(17) $(a+d)^2 < 4(ad - bc)$ which is equivalent with $(a-d)^2 < -4bc$

6.1.1 Markov Processes

(1) (a) Tuesday: $[0.4667, 0.3333, 0.2]^t$, Wednesday: $[0.4933, 0.3067, 0.2]^t$ (b) In two weeks the state vector is $[0.519993, 0.280007, 0.2]^t$ (c) The equilibrium distribution is $[0.52, 0.28, 0.2]^t$. After 50 days there is no noticeable change. (d) Let V_0 be the initial state, X the equilibrium state, and Y_1 and Y_2 bases for the other two eigenspaces. Then there are scalars b, c, and d such that $V_0 = bX + cY_1 + dY_2$. Then, since the entries of both sides total to 1, $a = 1$. The result follows by multiplying by A n times and taking the limit.(e) The entries of the equilibrium state vector give the expected fractions.

(2) (a)$[0.4167, 0.5833]^t$ (b) $[0.2195, 0.4146, 0.3659]^t$ (c) $[0.2000, 0.8000]^t$ (d) $[0.3469, 0.2449, 0.4082]^t$.

(3) (a) takes 4 products, (c) takes 5 products. (The answer in the back of the text is wrong.)

6.1. EIGENVECTORS

(5) Since the columns of P total to 1,
$$P = \begin{bmatrix} a & b \\ 1-a & 1-b \end{bmatrix}$$
Thus, $[1, -1]^t$ is an eigenvector with eigenvalue $a - b$.

(6) The probability is 0.7520. I am in a good mood 75% of the time.
$$P = \begin{bmatrix} 0.8 & 0.6 \\ 0.2 & 0.4 \end{bmatrix}$$

(7) (a) The equilibrium vector is $[0.2727, 0.1818, 0.5455]^t$ and the transition matrix is
$$\begin{bmatrix} .6 & .15 & .10 \\ .2 & .7 & .10 \\ .2 & .15 & .8 \end{bmatrix}$$

(9) The identity matrix has zero entries. The uniqueness of X fails. Since every vector V_0 satisfies $PV_0 = V_0$, it is true that $\lim_{n \to \infty} P^n V_0$ exists and equals a vector which is fixed by P.

(15) P has a zero entry. Consider $\lim_{n \to \infty} P^{2n}V$ and $\lim_{n \to \infty} P^{2n+1}V = \lim_{n \to \infty} P^{2n} PV$. From Theorem 1, (applied to P^2) both limits equal X. Hence $\lim_{n \to \infty} P^n V = X$.

(16) $P^t X = X$. Hence, $P^t - I$ is not invertible, showing that $\det(P - I) = \det(P^t - I) = 0$. Thus, there is a vector X_o such that $PX_o = X_o$. This is one part of Theorem 1.

(17) The i th entry of $P^t X$ is
$$p_{1i} x_1 + \ldots + p_{ni} x_n \leq p_{1i} M(X) + \ldots + p_{ni} M(X) = M(X)$$
The proof that each entry of $P^t X$ is greater than $m(X)$ is similar.

(a) Assume first that P has an eigenvalue $\lambda > 1$ and that X is a corresponding eigenvector so $P^t X = \lambda X$. Let x_i be the largest entry of X. If $x_i > 0$, then $\lambda x_i > x_i$. But this means that there is an entry of $P^t X$ which is larger than each entry of X, in contradiction to the previous argument. If $x_i < 0$, then all of the entries of X are negative. In this case, we replace X by $-X$ and repeat the argument to reach a contradiction. The assumption that $\lambda < -1$ reaches a similar contradiction by considering the smallest entry of X.

(b) Suppose that X is a vector such that $P^t X = X$. Let x_i be the largest entry of X. Then

$$x_i = p_{1i} x_1 + \ldots + p_{ni} x_n$$

If any one of the x_j is strictly less than x_i, then the right side of this equality is strictly less than the left. Thus, all of the x_j are equal proving that X is a multiple of $[1, 1, \ldots, 1]^t$.

(18) Let V_0 be a probability vector in R^n. We may write

$$V_0 = a_1 X_1 + \ldots + a_n X_n$$

where the X_i are eigenvectors for P corresponding to the eigenvalue λ_i and where $\lambda_1 = 1$ and $|\lambda_i| < 1$ for $i > 1$. Then, for all m,

$$P^m V_0 = a_1 X_1 + a_2 \lambda_2^m X_2 + \ldots + a_n \lambda_n^m X_n$$

As $m \to \infty$, this converges to $X = a_1 X_1$. Note that since V_0 is a probability vector, it follows that X is the unique probability vector in the span of $[1, 1, \ldots, 1]^t$.

(19) Proof: If -1 is an eigenvalue for P, then the $\lambda = 1$ eigenspace for P^2 is at least two dimensional, which contradicts Exercise 17.

6.2 Diagonalization

True-False Answers:
1. F, 2. F, 3. T, 4. F, 5. T, 6. F, 7. F (they are non-negative), 8. F.

(3) (a)

$$Q = \begin{bmatrix} 1 & 1 & 0 \\ 1 & 0 & 1 \\ -1 & 1 & 1 \end{bmatrix} \quad D = \begin{bmatrix} 6 & 0 & 0 \\ 0 & -3 & 0 \\ 0 & 0 & 0 \end{bmatrix} \quad A^n = Q \begin{bmatrix} 6^n & 0 & 0 \\ 0 & (-3)^n & 0 \\ 0 & 0 & 0 \end{bmatrix} Q^{-1}$$

(b)

$$Q = \begin{bmatrix} 1 & 0 & 1 \\ 0 & 1 & 1 \\ -2 & 1 & 1 \end{bmatrix} \quad D = \begin{bmatrix} 2 & 0 & 0 \\ 0 & 2 & 0 \\ 0 & 0 & 4 \end{bmatrix} \quad A^n = Q \begin{bmatrix} 2^n & 0 & 0 \\ 0 & 2^n & 0 \\ 0 & 0 & 4^n \end{bmatrix} Q^{-1}$$

6.2. DIAGONALIZATION

(4) Characteristic polynomials: (a) $\lambda^3 - 3\lambda^2$ (b) $\lambda^3 - 4\lambda^2 + 3$

Diagonalizations:

(a)
$$Q = \begin{bmatrix} 1 & -1 & -1 \\ 1 & 0 & 1 \\ 1 & 1 & 0 \end{bmatrix} \quad D = \begin{bmatrix} 0 & 0 & 0 \\ 0 & 0 & 0 \\ 0 & 0 & 3 \end{bmatrix}$$

(b)
$$Q = \begin{bmatrix} 1 & 1 & -1 \\ -1 & 2 & 1 \\ 1 & 1 & 0 \end{bmatrix} \quad D = \begin{bmatrix} 0 & 0 & 0 \\ 0 & 3 & 0 \\ 0 & 0 & 1 \end{bmatrix}$$

(5) (a) Not diagonalizable over R. (b) Not diagonalizable. (c)

$$Q = \begin{bmatrix} 1 & 1 & 0 & 0 \\ 1 & -1 & 0 & 0 \\ 0 & 0 & 2 & -1 \\ 0 & 0 & -1 & 1 \end{bmatrix} \quad D = \begin{bmatrix} 4 & 0 & 0 & 0 \\ 0 & -2 & 0 & 0 \\ 0 & 0 & -2 & 0 \\ 0 & 0 & 0 & -3 \end{bmatrix}$$

(6) (b) is not diagonalizable: The only eigenvalue is 4 and the eigenspace is spanned by $[1,3]^t$. For the other matrices we find

(a) $\quad Q = \begin{bmatrix} 1 & 1 \\ 3 & 2 \end{bmatrix} \quad D = \begin{bmatrix} 2 & 0 \\ 0 & -1 \end{bmatrix}$

(c) $\quad Q = \begin{bmatrix} 0 & 1 & 2 \\ 1 & 0 & 1 \\ 1 & 0 & 0 \end{bmatrix} \quad D = \begin{bmatrix} 3 & 0 & 0 \\ 0 & 1 & 0 \\ 0 & 0 & 2 \end{bmatrix}$

(d) $\quad Q = \begin{bmatrix} 0 & 1 & 0 \\ 1 & 0 & 1 \\ 1 & 3 & 0 \end{bmatrix} \quad D = \begin{bmatrix} 1 & 0 & 0 \\ 0 & 2 & 0 \\ 0 & 0 & 2 \end{bmatrix}$

(8) A is diagonalizable if and only if $A - 2I$ has rank 1 which is true if and only if $a = -7k$ and $b = ck$ where k is any scalar.

(9) Yes.

(10) $B = QCQ^{-1}$ where C has the square root of the eigenvalues as its diagonal entries. For Examples 1 and 2 we get, respectively,

$$\begin{bmatrix} \frac{\sqrt{2}}{2}+1 & -\frac{\sqrt{2}}{2}+1 \\ -\frac{\sqrt{2}}{2}+1 & \frac{\sqrt{2}}{2}+1 \end{bmatrix} \quad \begin{bmatrix} \sqrt{3} & -\frac{1}{2}+\frac{\sqrt{3}}{2} & 0 \\ 0 & 1 & 0 \\ -2+2\sqrt{3} & -1+\sqrt{3} & 1 \end{bmatrix}$$

(11) Write $A = QDQ^{-1}$ where D is diagonal. Then $D^2 = I$ which shows (formula 2) that $A^2 = I$.

(12) Write $A = QDQ^{-1}$ where D is diagonal. Then $D^2 = D$ which shows (formula 2) that $A^2 = A$.

(13) The proof is similar to that in Exercises 11 and 12.

(14) Write $A = QDQ^{-1}$ where D is diagonal. Each diagonal entry λ_i of D satisfies $q(\lambda_i) = 0$. Hence $q(D) = \mathbf{0}$, from which $q(A) = \mathbf{0}$ follows.

6.3 Complex Eigenvectors

True-False Answers:
  1. F, 2. F, 3. F, 4. T.

(1) $AB = \begin{bmatrix} -1+3i & -5+3i \\ -3+4i & -6 \end{bmatrix}, BA = \begin{bmatrix} 7-i & 2+9i \\ 1+11i & -14+4i \end{bmatrix}$

(2) $A = QDQ^{-1}$ where

$$Q = \begin{bmatrix} 1 & 1 \\ -i/2 & ,i/2 \end{bmatrix} \quad D = \begin{bmatrix} 1+2i & 0 \\ 0 & 1-2i \end{bmatrix}$$

Also

$$(1\pm 2i)^{20} = (\sqrt{5})^{20} \left(\cos(20(1.10715))\pm\sin(20(1.10715))\right) = -5^{10}(.98850\pm.15124i)$$

Hence

$$A^{20} = 5^{10} \begin{bmatrix} -0.9884965888 & 0.3024863232 \\ -0.07562158080 & -0.9884965888 \end{bmatrix}$$

6.4. MATRIX OF A LINEAR TRANSFORMATION

(4) The eigenvalues are $2 \pm 3i$

$$A = \begin{bmatrix} 2 & -3 \\ 3 & 2 \end{bmatrix}$$

(3) $-7, 6.5 \pm \frac{3\sqrt{3}}{2}i$

(5) $M = \begin{bmatrix} a & -b \\ b & a \end{bmatrix}$. Eigenvalues $a \pm bi$

(6) The eigenvalues are $\lambda_\pm = \cos\theta \pm i\sin\theta$ coresponding, respectively, to $[1, \pm i]^t$.

(17) (a) $3 + 20i$

(21) They satisfy $||U_i|| = 1$ and $<U_i, U_j> = 0$ if $i \neq j$. For the proof, see the argument for Proposition 1 in Section 4.4.

6.4 Matrix of a Linear Transformation

True-False Answers: 1. T, 2. T.

(1) $\begin{bmatrix} 3 & 1 & 0 \\ 2 & 0 & 2 \\ 1 & 1 & 1 \end{bmatrix}$

(2)
$$M = \begin{bmatrix} 0 & 0 & 0 \\ 2 & 0 & 0 \\ 5 & 4 & 0 \end{bmatrix}$$

(3) (i) Note that $AQ_i = \sum_j m_{ji} Q_j$. Thus, the answer to (i) may be read off of M.

(a) $M = \begin{bmatrix} 3 & 1 \\ 0 & 3 \end{bmatrix}$

(b) $M = \begin{bmatrix} 3 & 0 & 0 \\ 6 & 3 & 0 \\ 3 & 0 & 6 \end{bmatrix}$

(4) (a) $Q = \begin{bmatrix} 1 & -2 \\ 0 & -2 \end{bmatrix}$ $M = \begin{bmatrix} 3 & 0 \\ 1 & 3 \end{bmatrix}$

(b) $Q = \begin{bmatrix} 1 & -9 \\ 0 & -3 \end{bmatrix}$ $M = \begin{bmatrix} -2 & 0 \\ 1 & -2 \end{bmatrix}$

(5)
$$Q = \begin{bmatrix} 0 & -4 & 2 \\ 0 & 1 & 0 \\ 1 & 0 & 0 \end{bmatrix}$$

(6)
$$M^k = \begin{bmatrix} 1 & 0 \\ k & 0 \end{bmatrix} \quad A^k = \begin{bmatrix} 1+\frac{k}{2} & -\frac{k}{2} \\ \frac{k}{2} & -\frac{k}{2}+1 \end{bmatrix}$$

(7)
$$M^k = 3^k \begin{bmatrix} 1 & 0 & 0 \\ \frac{k}{3} & 1 & 0 \\ 0 & 0 & \left(\frac{5}{3}\right)^k \end{bmatrix} \quad A^k = \begin{bmatrix} 3^k & k3^k & \frac{3(5^k)-2k3^k-3^{k+1}}{4} \\ 0 & 4(3^k) & \frac{5^k-3^k}{2} \\ 0 & 0 & 5^k \end{bmatrix}$$

(8)
$$M = \begin{bmatrix} 0 & 0 & 0 \\ 1 & 0 & 0 \\ 0 & 1 & 0 \end{bmatrix}$$

(9)
$$M = \begin{bmatrix} 0 & 0 & 0 & \cdots & 0 \\ 1 & 0 & 0 & \cdots & 0 \\ 0 & 1 & 0 & \cdots & 0 \\ \vdots & \vdots & \ddots & \ddots & \vdots \\ 0 & 0 & \cdots & 1 & 0 \end{bmatrix}$$

(11) (a) $M = \begin{bmatrix} 0 & 2 \\ 1 & -1 \end{bmatrix}$ (b) $X' = [\frac{5}{2}, \frac{1}{2}]^t$, $X'' = [1, 2]^t$.

(13) (a) $M = \begin{bmatrix} 0 & 1 & 0 \\ 0 & 0 & 2 \\ 0 & 0 & 0 \end{bmatrix}$, nullspace is spanned by $[1, 0, 0]^t$ which corresponds to the constant function $p_0(x) = 1$. The image is spanned by $[0, 2, 0]^t$ and $[1, 0, 0]^t$ which correspond to the functions $p_0(x) = 1$ and $p_1(x) = 2x$. p_0 is the derivative of $f(x) = x$ and p_1 is the derivative of $f(x) = x^2$.

(14) (a) $M = \begin{bmatrix} 0 & 1 & 0 & 0 \\ 0 & 0 & 2 & 0 \\ 0 & 0 & 0 & 3 \\ 0 & 0 & 0 & 0 \end{bmatrix}$, nullspace is spanned by $[1, 0, 0, 0]^t$ which corresponds to the constant function $p_0(x) = 1$. The image is spanned

6.5. ORTHOGONAL DIAGONALIZATION

by $[1,0,0,0]^t$, $[0,2,0,0]$, and $[0,0,3,0]^t$ which correspond to the functions 1, $2x$, and $3x^2$ which are, respectively, the derivatives of x, x^2 and x^3.

(15) $M = \begin{bmatrix} 0 & 1 & 1 \\ 0 & 0 & 2 \\ 0 & 0 & 0 \end{bmatrix}$

(16) $M = \begin{bmatrix} 0 & 1 & 1 & 1 \\ 0 & 0 & 2 & 3 \\ 0 & 0 & 0 & 3 \\ 0 & 0 & 0 & 0 \end{bmatrix}$

(17) The matrix is the same as in Exercise 13. The image is spanned by $2p_1$ and p_0.

(18) The matrix is the same as in Exercise 14. The image is spanned by $3p_2$, $2p_1$, and p_0.

6.5 Orthogonal Diagonalization

True-False Answers:

1. F, 2. F (assuming that the matrix is real), 3. F, 4. T, 5. F, 6. T, 7. F.

(1) The only matrices that describe $X^t A X = d$ (with the same d) are as follows. If d is allowed to change, then any non-zero scalar multiple of this matrix also works.

$$\begin{bmatrix} 3 & a \\ 2-a & 3 \end{bmatrix}$$

(2) Equations: (a) $-4(x')^2 + 6(y')^2 = 1$ (b) $2(x'_2)^2 + 4(x'_3)^2 = 4$ (c) $28(y')^2 + 42(z')^2 = 7$. Orthogonal matrices (the columns are the bases):

$$\begin{bmatrix} \frac{\sqrt{5}}{5} & 2\frac{\sqrt{5}}{5} \\ -2\frac{\sqrt{5}}{5} & \frac{\sqrt{5}}{5} \end{bmatrix} \quad \begin{bmatrix} 1/2 & \frac{\sqrt{2}}{2} & 1/2 \\ -\frac{\sqrt{2}}{2} & 0 & \frac{\sqrt{2}}{2} \\ 1/2 & -\frac{\sqrt{2}}{2} & 1/2 \end{bmatrix}$$

$$\begin{bmatrix} 2/7 & 3/7 & 6/7 \\ 6/7 & 2/7 & -3/7 \\ 3/7 & -6/7 & 2/7 \end{bmatrix}$$

(3) If both eigenvalues are ≤ 0, then there is no graph. If exactly one is zero, then the figure is a pair of parallel lines. If both are positive, it is an ellipse and if they are mixed sign, it is an hyperbola.

(4) The curves (a)-(c) are, respectively, an ellipse, an hyperbola and a pair of lines and the normal forms are

$$\left(3+\sqrt{2}\right)x^2 + \left(3-\sqrt{2}\right)y^2 = 1$$
$$\left(3+\sqrt{17}\right)x^2 + \left(3-\sqrt{17}\right)y^2 = 1$$
$$10\,y^2 = 1$$

(5)
$$1/20\,x^2 - 1/30\,xy + \frac{29}{720}y^2 = 1$$

(6) The hyperbola whose formula in rotated coordinates is $ax^2 - by^2 = 1$ has the following formula in natural coordinates

$$\left(\frac{9}{25}a - \frac{16}{25}b\right)x^2 + \left(\frac{24}{25}a + \frac{24}{25}b\right)yx + \left(\frac{16}{25}a - \frac{9}{25}b\right)y^2$$

(8) The $\lambda = 2$ eigenspace is 3 dimensional. P will depend on how the students order the basis. One answer is, where the last column is a basis for the $\lambda = 9$ eigenspace.

$$\begin{bmatrix} -2\frac{\sqrt{5}}{5} & \frac{\sqrt{30}}{30} & -1/42\sqrt{42} & \frac{\sqrt{7}}{7} \\ 0 & 0 & \frac{\sqrt{42}}{7} & 1/7\sqrt{7} \\ 0 & \frac{\sqrt{30}}{6} & -\frac{\sqrt{42}}{42} & \frac{\sqrt{7}}{7} \\ \frac{\sqrt{5}}{5} & -\frac{\sqrt{30}}{15} & -\frac{\sqrt{42}}{21} & 2\frac{\sqrt{7}}{7} \end{bmatrix}$$

(12) An eigenvector corresponding to the eigenvalue -4 will suffice. Thus, let $X = [-\frac{\sqrt{2}}{2}, 1, \frac{\sqrt{2}}{2}]^t$.

Mathematica On Line

On Line
1.2 Matrices

■ Discussion

■ Linear Combinations of Vectors

Our goal in this dicussion is to plot some elements of the span of the vectors $a = \{1, 1\}$ and $b = \{1, 3\}$. We'll begin by seeing how to enter vectors into the computer. *Mathematica* uses lists as its basic objects for representing vectors. Lists must begin and end with curly braces "{" and "}". Round parentheses "(" and ")" and square brackets "[" and "]" will not indicate a list and will cause significant syntax errors. Here are the vectors a and b. If you are reading the electronic version of this notebook you should enter these expressions into *Mathematica*. Do this by clicking the mouse somewhere in the expression (or in the cell bracket to the right of the expression) and while pressing the "shift" key, press the "return" key (on a Macintosh) or the "enter" key (on an IBM). Note that the "return" or "enter" keys, by themselves, give you a new line on which to continue typing your expression; they *do not* evaluate the expression.

```
In[1]:= Clear[a, b]
        a = {1, 1}
        b = {1, 3}
Out[2]= {1, 1}
Out[3]= {1, 3}
```

Next we construct a few elements of the span of a and b. First, we will enter $2a + b$.

```
In[4]:= 2 a + b
Out[4]= {3, 5}
```

If we enter -5 a + 7 b, *Mathematica* will compute another element of the span.

In[5]:= **-5 a + 7 b**

Out[5]= {2, 16}

Thus, the vectors {3, 5} and {2, 16} both belong to the span of *a* and *b*.

■ Random Linear Combinations of Vectors

We can get *Mathematica* to automatically generate elements of the span. We will use the function *Random* which produces random numbers between 0 and 1. Here is a list of three random numbers.

In[6]:= **{Random[], Random[], Random[]}**

Out[6]= {0.986394, 0.417517, 0.374104}

The following expression will generate a random linear combination, *c*, of *a* and *b*.

In[7]:= **Clear[c]**
       c = Random[] a + Random[] b

Out[8]= {0.469844, 0.922389}

To see more random combinations of *a* and *b* just enter the expression above by clicking the mouse somewhere in the expression and while pressing the "shift" key, press the "return" key or the "enter" key.

Rather than just repeating the command to generate a random combination of vectors *a* and *b*, we can create a list of them, *randomList*, using the function *Table*.

In[9]:= **Clear[randomList]**
       randomList = Table[Random[] a + Random[] b, {k, 1, 5}]

Out[10]= {{1.32202, 2.53207}, {0.439695, 0.581714},
    {1.09015, 2.39178}, {1.38835, 2.51667},
    {1.46577, 2.60103}}

1.2 Mathematica On Line.nb

If we want a larger list but don't want to look at the specific output we can suppress printing the output by using a semi-colon, ";", at the end of the expression. We define *randomList2* to be a list of 50 random combinations of *a* and *b*.

```
In[11]:= Clear[randomList2]
         randomList2 =
           Table[Random[] a + Random[] b, {k, 1, 50}];
```

■ Plotting Linear Combinations of Vectors

We can plot our list of linear combinations, *randomList2*, by using the *Mathematica* function *ListPlot*.

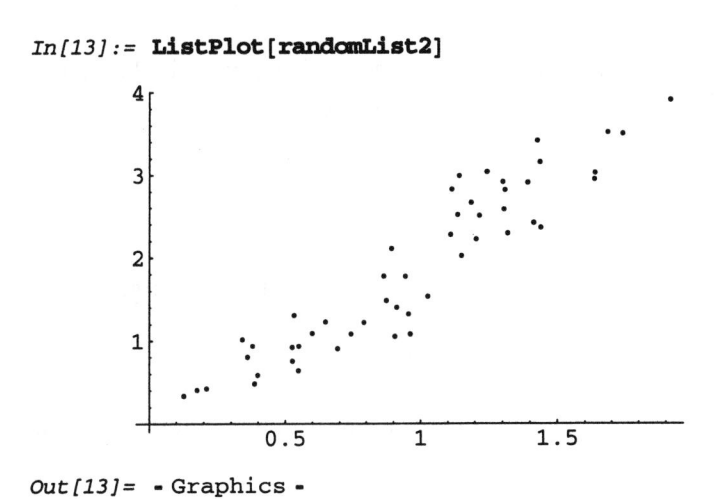

```
In[13]:= ListPlot[randomList2]
```

Out[13]= - Graphics -

■ Finding Components of Vectors

Mathematica lets us extract components from vectors. First we define a vector *v*.

```
In[14]:= Clear[v]
         v = {14, -7, Pi}
```
Out[15]= {14, -7, π}

The first component of *v* can be found by the following expression which gives the first item in the list *v*.

```
In[16]:= v[[1]]
```

```
Out[16]= 14
```

Similarly, the second and third components can be found by the following expressions.

```
In[17]:= v[[2]]
```

```
Out[17]= -7
```

```
In[18]:= v[[3]]
```

$Out[18]= \pi$

(The double square brackets, "[[", and "]]", are required because single square brackets, "[", and "]", are reserved for use with functions.)

■ Exercises

1. Plot the points {1, 1}, {1, -1}, {-1, 1}, and {-1, -1}, all on the same figure.

2. Enter the vectors *a* and *b* from the discussion above.

(a) Use *Mathematica* to compute several different linear combinations of them. (Reader's choice.)

(b) Use *c = Random[] a + Random[] b* to create several "random" linear combinations of *a* and *b*.

(c) Plot enough points in the span of *a* and *b* to get a discernible geometric figure. You will want to use the function *Table* here. Use the semi-colon, ";", to suppress your written output. You may need to plot as many as 100 or 500 points. What kind of geometric figure do these points seem to form? What are the coordinates of the vertices?

(d) The plot in part (c) is only part of the span. To see more of the span, enter the following expression. The *PlotStyle* option specifies a color with red, green, and blue components, each between 0 and 1. In the expression below the color is red. Experiment with other colors until you find one you like by changing the 0's and 1's to other numbers between 0 and 1. After you have edited the expression you may enter it again by using the "shift-enter" or the "shift-return" combinations.

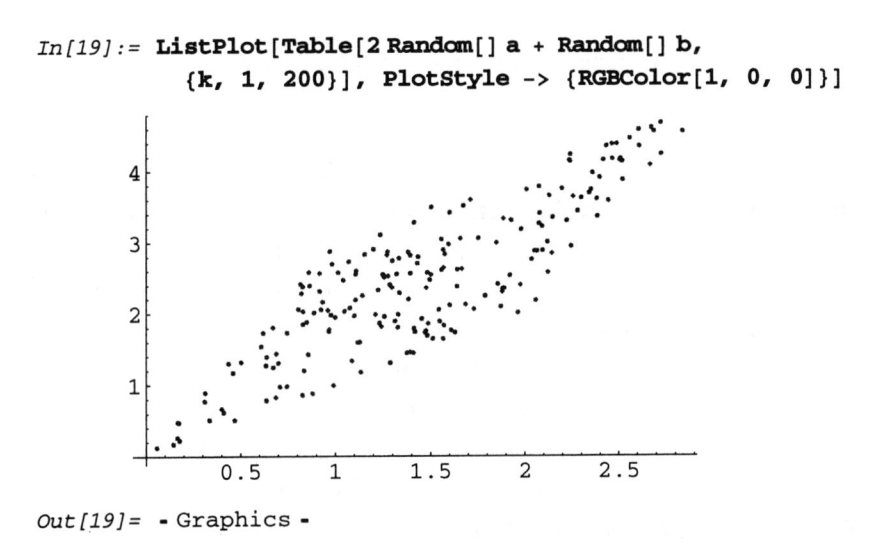

```
In[19]:= ListPlot[Table[2 Random[] a + Random[] b,
         {k, 1, 200}], PlotStyle -> {RGBColor[1, 0, 0]}]
```

Out[19]= - Graphics -

3. Describe in words the set of points $s\,a + t\,b$ for $-2 \leq s \leq 2$ and $-2 \leq t \leq 2$. Create a *Mathematica* plot that shows this set reasonably well. (Hint: *Random[]* gives random numbers in the interval [0, 1] and *Random[{a,b}]* gives random numbers in the interval [a, b].)

4. In exercise 9 of Section 1.2, it was stated that each element, $\{x, y, z\}$, of the span of $X = \{-1, 1, -1\}$ and $Y = \{-1, 3, 2\}$ satisfies $5x + 3y - 2z = 0$. Clearly, X and Y satisfy this condition:

```
In[20]:= Clear[X, Y]
        X = {-1, 1, -1};
        Y = {-1, 3, 2};
        5 X[[1]] + 3 X[[2]] - 2 X[[3]]

Out[21]= 0

In[22]:= 5 Y[[1]] + 3 Y[[2]] - 2 Y[[3]]

Out[22]= 0
```

(a) Check this condition by generating a random vector c in the span of X and Y and computing $5\ c[[1]] + 3\ c[[2]] - 2\ c[[3]]$.

(b) Plot a few hundred elements of this span in R^3. Here is a plot of 100 points to help you get started. The only thing that will need to be changed is the range for the index k that determines the number of elements that are generated by *Table*. (The functions here may be confusing but don't let that bother you. At this point you don't need to understand the individual commands, just that they are producing a graph of random combinations of X and Y. For those who might be interested: the *Point* function changes the vectors to graphics primitives, the *Graphics3D* function creates graphics images from the primitives and *Show* displays the graphics images. If you are still thirsting for more, go directly to Spephen Wolfram's *The Mathematica Book*, third edition.)

1.2 Mathematica On Line.nb   81

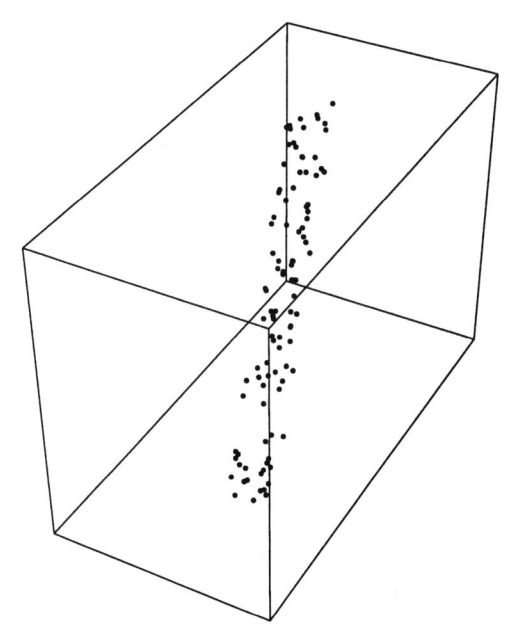

```
In[23]:= Show[Graphics3D[
          Table[Point[Random[] X + Random[] Y],
           {k, 1, 100}]]]
```

Out[23]= - Graphics3D -

(c) Describe the geometric figure you have obtained.

On Line
1.3 Systems

■ Exercises

1. Solve by hand the following equation:

$$x_1 + 2x_2 + 3x_3 = 4. \qquad (4)$$

You should obtain the general solution

$$x = \{4, 0, 0\} + s\{-2, 1, 0\} + t\{-3, 0, 1\}.$$

(a) Use the *Mathematica* function *Random[ ]* to create a "random" element c of the span of the vectors $v = \{-2, 1, 0\}$ and $w = \{-3, 0, 1\}$. (See the On Line portion of Section 1.2.)

(b) Let $x = \{4, 0, 0\} + c$. Check by direct substitution that x solves equation (4). This demonstrates that the general solution to our system is the span of the spanning vectors translated by the translation vector.

(c) Substitute c into equation (4) and describe what you get. Try substituting some other random linear combinations of v and w into this equation. What do you conjecture? Can you prove it?

2. Repeat the above sequence (a) - (c) for the system (U) which was solved in Section 1.3. Specifically, in (a) you will create a random element c in the span of the spanning vectors $v = \{-2, 1, 1, 0\}$ and $w = \{-2, 1, 0, 1\}$. In (b) you will substitute $c + \{-1, 1, 0, 0\}$ into *each* equation in the system and in (c) you will substitue c into *each* equation in the system.

Does the conjecture you made in Exercise (1) above still hold? Can you prove it?

On Line
1.4 Gaussian Elimination

■ Discussion

■ Defining Matrices

We saw in Section 1.2 that *Mathematica* represents vectors as lists. In this section we will examine *Mathematica*'s representation of matrices as lists of lists. Here is an example. (We start by clearing the variable *A*. Generally, it is always a good idea to clear variables before defining them.)

In[1]:= **Clear[A]**
     A = {{1, 2, 3}, {4, 5, 6}}

Out[2]= {{1, 2, 3}, {4, 5, 6}}

To see the matrix *A* in the traditional rectangular form we use the function *MatrixForm*.

In[3]:= **MatrixForm[A]**

Out[3]//MatrixForm=
$$\begin{pmatrix} 1 & 2 & 3 \\ 4 & 5 & 6 \end{pmatrix}$$

Another way to use *MatrixForm* is with "postfix" notation. This method is nice because the function *MatrixForm* gets pushed into the background.

In[4]:= **A // MatrixForm**

Out[4]//MatrixForm=
$$\begin{pmatrix} 1 & 2 & 3 \\ 4 & 5 & 6 \end{pmatrix}$$

1.4 Mathematica On Line.nb

■ Extracting Rows and Elements of Matrices

The structure of *A* as a lists of lists means that the first list of *A* is the first row of *A* and the second list of *A* is the second row of *A*. Thus we can obtain the first row of *A* as follows:

In[5]:= **A[[1]]**

Out[5]= {1, 2, 3}

Similarly, we have the second row of *A*:

In[6]:= **A[[2]]**

Out[6]= {4, 5, 6}

We can also find an individual entry of *A*. Here is the element in the 2nd row and 3rd column.

In[7]:= **A[[2, 3]]**

Out[7]= 6

■ Row Reduction

Mathematica has a function, *RowReduce*, which computes the reduced echelon form of any given matrix. One begins by entering the matrix into *Mathematica*. Here is a new matrix *A*.

```
In[8]:= Clear[A]
       A = {{1, -1, 1, 3, 0, 6},
            {2, -2, 2, 6, 0, 7},
            {-1, 1, 1, -1, -2, 1},
            {4, -4, 1, 9, 3, 6}};
       MatrixForm[A]
```

Out[9]//MatrixForm=
$$\begin{pmatrix} 1 & -1 & 1 & 3 & 0 & 6 \\ 2 & -2 & 2 & 6 & 0 & 7 \\ -1 & 1 & 1 & -1 & -2 & 1 \\ 4 & -4 & 1 & 9 & 3 & 6 \end{pmatrix}$$

We can row reduce A as follows:

```
In[10]:= RowReduce[A] // MatrixForm
```

Out[10]//MatrixForm=
$$\begin{pmatrix} 1 & -1 & 0 & 2 & 1 & 0 \\ 0 & 0 & 1 & 1 & -1 & 0 \\ 0 & 0 & 0 & 0 & 0 & 1 \\ 0 & 0 & 0 & 0 & 0 & 0 \end{pmatrix}$$

■ Exercises

1. Use *RowReduce* to find all solutions to the system in Exercise 5(g) from Section 1.3.

2. We said that the rank of a system is the number of equations left after eliminating dependent equations. We commented (but certainly did not prove) that this number does not depend on which equations we kept and which we eliminated. It seems natural to suppose that the rows in the row reduced form of a matrix represent independent equations and hence that the rank should also be computable as the number of non-zero rows in the row reduced form of the matrix. We can check this conjecture experimentally using *Mathematica*. Create four 4 by 5 matrices A_1, A_2, A_3, and A_4, having respective ranks of 1, 2, 3, and 4. Design your matrices so that none of their entries are zero. Then compute their rank by using *RowReduce*.

1.4 Mathematica On Line.nb 87

3. For each of the four matrices from the previous exercise, use *RowReduce* to row reduce their transpose. The *Mathematica* expression for the transpose of a matrix A is *Transpose[A]*. How does the rank of each matrix compare with that of its transpose?

4. Here are three vectors, x, y, z.

```
In[11]:= Clear[x, y, z]
         x = {-5, 23, -41, 46, 8}
         y = {6, 1, -8, 2, 10}
         z = {-5, 12, -19, 24, 1}

Out[12]= {-5, 23, -41, 46, 8}

Out[13]= {6, 1, -8, 2, 10}

Out[14]= {-5, 12, -19, 24, 1}
```

Here are two additional vectors, u and v.

```
In[15]:= Clear[u, v]
         u = {-5, 23, -4, 46, 8}
         v = {22, 0, -22, 0, 34}

Out[16]= {-5, 23, -4, 46, 8}

Out[17]= {22, 0, -22, 0, 34}
```

(a) Determine which of the vectors u and v is in the span of x, y, and z by using *RowReduce* to solve a system of equations as in Exercise 6 of Section 1.4. Here is how we can create a matrix with the vectors x, y, z, and u for rows.

```
In[18]:= Clear[A]
         A = {x, y, z, u};
         MatrixForm[A]

Out[19]//MatrixForm=
         ⎛ -5   23   -41   46    8 ⎞
         ⎜  6    1    -8    2   10 ⎟
         ⎜ -5   12   -19   24    1 ⎟
         ⎝ -5   23    -4   46    8 ⎠
```

(b) Imagine that you are the head of an engineering group and that you have a computer technician working for you who knows *absolutely nothing* about linear algebra, other than how to enter matrices and commands into *Mathematica*. You need to tell your technician how to do problems similar to part (a) above. Specifically, you will give him/her an initial set of three vectors x, y, and z from R^3. You will then provide an additional vector u and you want the technician to determine whether u is in the span of x, y, and z.

Write a brief set of instructions which will tell your technician how to do this job. Be as explicit as possible. Remember that the technician cannot do lienar algebra! You must provide instructions on how to construct the necessary matrices, what to do with them and how to interpret the answers. The final "output" to you should be a simple "Yes" or "No". You don't want to see the matrices!

On Line
1.5 Column Space and Null Space

■ Discussion

■ Multiplying Matrices and Vectors

Mathematica is very good at multiplying matrices and vectors. For example, here is a matrix A and a vector x.

```
In[1]:= Clear[A, x]
        A = {{1, 2, 1, 3},
             {-5, 7, 2, 2},
             {13, 4, 4, 3}};
        MatrixForm[A]
        x = {1, 3, -2, 4}
```

$Out[2]//MatrixForm=$
$$\begin{pmatrix} 1 & 2 & 1 & 3 \\ -5 & 7 & 2 & 2 \\ 13 & 4 & 4 & 3 \end{pmatrix}$$

$Out[3]= \{1, 3, -2, 4\}$

Now we can compute the product $b = A \cdot x$.

```
In[4]:= Clear[b]
        b = A . x
```

$Out[5]= \{17, 20, 29\}$

Caution: Notice the period, ".", between A and x. Without this period you will not be able to correctly compute the product.

Because of the structure of vectors as lists and mastrices as lists of lists, *Mathematica* is able to distinguish vectors from matrices. This means that we don't have to worry about whether a vector is in "row form" or "column form" before it is multiplied by a matrix.

■ The Matrix Manipulation Package

For manipulating matrices we can use functions that are not built into *Mathematica* but instead are contained in a special package. To activate this package we need to evaluate the following expression.

In[6]:= **<< LinearAlgebra`MatrixManipulation`**

■ Obtaining Columns from a Matrix

Obtaining columns of a matrix is equivalent to obtaining rows of the transpose of a matrix. Here is an example. We write out *A* in matrix form so that we can check our answers easily.

In[7]:= **MatrixForm[A]**

Out[7]//MatrixForm=
$$\begin{pmatrix} 1 & 2 & 1 & 3 \\ -5 & 7 & 2 & 2 \\ 13 & 4 & 4 & 3 \end{pmatrix}$$

Here is the second column of *A* expressed as a vector.

In[8]:= **Clear[col2]**
        col2 = Transpose[A][[2]]

Out[9]= {2, 7, 4}

Using the *MatrixManipulation* package we can also obtain columns of a matrix. Here is the fourth column of *A* expressed as a "column Matrix."

In[10]:= **Clear[col4]**
         col4 = TakeColumns[A, {4}]

Out[11]= {{3}, {2}, {3}}

1.5 Mathematica On Line.nb

Notice the difference between these two results. In finding col_2 we have the result in vector form and in finding col_4 we have the result in the form of a 3 × 1 matrix. We can go from the vector form of col_2 to the matrix form with the *Partition* function.

In[12]:= **Partition[col2, 1]**

Out[12]= {{2}, {7}, {4}}

We can also change the form of col_4 from a matrix to a vector using the *Flatten* function.

In[13]:= **Flatten[col4]**

Out[13]= {3, 2, 3}

■ Building Augmented Matrices

If we wish to build an augmented matrix of the form [A b] we can use the function *AppendRows* from the *MatrixManipulation* package. We must first change b from a vector to a column matrix and then apply the *AppendRows* function. We can do both of these tasks at the same time.

In[14]:= **Clear[augmentedAb]**
        augmentedAb = AppendRows[A, Partition[b, 1]];
        MatrixForm[augmentedAb]

Out[15]//MatrixForm=
$$\begin{pmatrix} 1 & 2 & 1 & 3 & 17 \\ -5 & 7 & 2 & 2 & 20 \\ 13 & 4 & 4 & 3 & 29 \end{pmatrix}$$

■ Exercises

1. Check the value of A . x computed above by asking *Mathematica* to compute $A_1 + 3 A_2 - 2 A_3 + 4 A_4$ where A_i is the ith column of A.

2. For the matrix A and vector b above, find the general solution to the system $Ax = b$ using *RowReduce* on the augmented matrix [A b]. Express your solution in parametric form. For which value of the parameter do you obtain x above?

3. Here is another way to find the general solution to the system in the above problem.

In[16]:= **NullSpace[A]**

Out[16]= {{-43, -99, 238, 1}}

The function *NullSpace* returns a list of basis vectors for the null space of A. We can extract a vector from this null spacce as follows:

In[17]:= **z = NullSpace[A][[1]]**

Out[17]= {-43, -99, 238, 1}

We can check this by computing $A \cdot z$.

In[18]:= **A . z**

Out[18]= {0, 0, 0}

We can also comput $A \cdot (x + 20\,z)$.

In[19]:= **A . (x + 20 z)**

Out[19]= {17, 20, 29}

Compare this with $A \cdot x$.

In[20]:= **A . x**

Out[20]= {17, 20, 29}

This example demonstrates the Translation Theorem, in that it shows that the general solution is any one solution plus the nullspace.

4. Here is a matrix A and a vector b.

1.5 Mathematica On Line.nb

```
In[21]:= Clear[A]
         A = {{17, -6, 13, 27, 64, 19},
              {4, -6, -33, 25, 7, 9},
              {55, -24, 6, 106, 199, 66},
              {89, -36, 32, 160, 327, 104}};
         A // MatrixForm
```

$$Out[22]//MatrixForm = \begin{pmatrix} 17 & -6 & 13 & 27 & 64 & 19 \\ 4 & -6 & -33 & 25 & 7 & 9 \\ 55 & -24 & 6 & 106 & 199 & 66 \\ 89 & -36 & 32 & 160 & 327 & 104 \end{pmatrix}$$

```
In[23]:= Clear[b]
         b = {17, 4, 55, 89}
Out[24]= {17, 4, 55, 89}
```

(a) Compute the rank of *A* by using *RowReduce*. How many free variables with the system $A \cdot x = 0$ have?

(b) How many spanning vectors will the nullspace for *A* have?

(c) Use the expression W = *NullSpace[A]* to find a spanning set for the nullspace of *A*. (The rows of *W* will be the desired spanning set.)

(d) Find, by using *RowReduce* on the augmented matrix [*A* *b*], a vector *x* such that $A \cdot x = b$.

(e) Recall that the ith row of *W* is given by *W[[i]]*. Check that

$$x + 2 W[[1]] - 3 W[[2]] + 7 W[[3]] - 9 W[[4]]$$

is a solution to $A \cdot x = b$.

(f) I claim that the general solution to $A \cdot x = b$ is x_0 + *Transpose[W]* · *y* where *y* is an element of R^4. Explain.

(g) Use *RowReduce* on the augmented matrix [*A* *0*] to find a basis for the nullspace of *A*. Your answer will be different from the answer that *NullSpace* gives. Demonstrate that the two answers are really equivalent by showing that the rows of *W* are multiples of your basis vectors.

On Line
1.5.1 Predator-Prey Problems
Linear Dynamical Systems

■ **Discussion**

Rabbits are pleasant little animals which, when left to their own devices, like nothing better than eating and producing more rabbits. In the presence of sufficient food and lacking natural competitors, we might guess that each spring the number of rabbits would increase by a fixed percentage, say 10%. Thus, if the initial rabbit population is R_0, then next spring there will be $R_1 = (1.1) R_0$ rabbits. In two years, the rabbit population will be $R_2 = (1.1)^2 R_0$. The number of rabbits after n years will be

$$R_n = (1.1)^n R_0.$$

The rabbit population grows without bound. Let us assume that R_n is measured in units of hundreds of rabbits.

Let us suppose now that we introduce a small number of mountain lions into our environment in order to keep the rabbit population under control. The rate of growth of the mountain lions will depend on both the number of mountain lions and the number of rabbits. We shall assume that the growth of the mountain lion population is governed by the equation

$$M_{n+1} = (0.4) R_n + (0.7) M_n. \qquad (1)$$

(Note that if $R_n = 0$, then the above equation says that the mountain lion population *decreases* by 30% each year. The mountain lions need the rabbits as food to survive.)

On the other hand, each year, the number of rabbits will be reduced by the number of rabbits which were eaten. We will assume that this is proportional to the number of mountain lions. Specifically we assume that

$$R_{n+1} = (1.1)R_n - (0.1)M_n \qquad (2)$$

We combine thse two equations into a single matrix equation as

$$\begin{pmatrix} R_{n+1} \\ M_{n+1} \end{pmatrix} = \begin{pmatrix} (1.1)R_n - (0.1)M_n \\ (0.4)R_n + (0.7)M_n \end{pmatrix} = \begin{pmatrix} 1.1 & -0.1 \\ 0.4 & 0.7 \end{pmatrix} \begin{pmatrix} R_n \\ M_n \end{pmatrix}.$$

It follows that each year's population is determined by multiplying the previous year's population by the 2×2 matrix above! Specifically, if T is the 2×2 matrix above and we set $P_n = \begin{pmatrix} R_n \\ M_n \end{pmatrix}$, then the above formula says that

$$P_{n+1} = TP_n.$$

Thus, for example, if our initial population is 10 hundred rabbits and 5 hundred mountain lions, then $P_1 = \begin{pmatrix} 10 \\ 5 \end{pmatrix}$. In year two, the population is

$$P_2 = \begin{pmatrix} 1.1 & -0.1 \\ 0.4 & 0.7 \end{pmatrix} \begin{pmatrix} 10 \\ 5 \end{pmatrix} = \begin{pmatrix} 10.5 \\ 7.5 \end{pmatrix}$$

In the third year, we have

$$P_3 = \begin{pmatrix} 1.1 & -0.1 \\ 0.4 & 0.7 \end{pmatrix} \begin{pmatrix} 10.5 \\ 7.5 \end{pmatrix} = \begin{pmatrix} 10.80 \\ 9.45 \end{pmatrix}.$$

It seems that the rabbit population is growing slowly and the mountain lion population is growing much more rapidly. We can see this clearly using *Mathematica*. We start with the initial population.

```
In[1]:= population[0] = {10, 5}
Out[1]= {10, 5}
```

1.5.1 Mathematica On Line.nb

Next, we define the matrix *T*.

In[2]:= `matrixT = {{1.1, -0.1}, {0.4, 0.7}}`

Out[2]= `{{1.1, -0.1}, {0.4, 0.7}}`

Finally, we define the population vector by recursion.

In[3]:= `population[n_] :=`
         `population[n] = matrixT . population[n-1]`

We are now ready to find the populations for the first 10 years.

In[4]:= `tenYears = Table[population[n], {n, 1, 10}]`

Out[4]= `{{10.5, 7.5}, {10.8, 9.45}, {10.935, 10.935}, {10.935, 12.0285},`
         `{10.8257, 12.794}, {10.6288, 13.285}, {10.3631, 13.5517},`
         `{10.0442, 13.6315}, {9.68551, 13.5597}, {9.29809, 13.365}}`

We can plot the populations. (The *PlotRange* option specifies the portion of the axes to be shown and the *AxesLabel* option specifies the labels for the axes.)

In[5]:= `ListPlot[tenYears, PlotRange -> {{9, 11}, {6, 14}},`
         `AxesLabel -> {"Rabbits", "Mountain Lions"},`
         `Ticks ->`
         `{{9, 9.5, 10, 10.5, 11}, {7, 8, 9, 10, 11, 12, 13, 14}}]`

Out[5]= `- Graphics`

■ Exercises

1. Plot enough population vectors to get a good idea of the long term population distribution of rabbits and mountain lions. (It should take 30 or more points.) Approximately what is the maximum mountain lion population? What is the maximum rabbit population? What does the model predict for the future of the rabbits and mountain lions?

2. Suppose that we begin with 10,000 rabbits and 200 mountain lions. What is the mountain lion population in year 2? Is this rate of growth reasonable?

3. The -0.1 entry in T is called the "kill rate." Try increasing this to -0.4. Begin with $P_1 = \{10, 5\}$. Plot a large number of points (50 or so). You should get a very "pretty" graph! Your pretty graph, however, is physically impossible! Explain.

4. The last two exercises suggest some problems with our model. Actually, this model is seriously flawed. In this exercise, we would like you to discuss some of the flaws in this model. For your discussion consider the following:

(a) Suppose that we initially have 10 hundred mountain lions in our population. According to equation (2), how many rabbits get eaten by mountain lions in the first year? (Be careful. The answer is not 100. Remember that eaten rabbits don't have offspring.) Note that this number is *independent of the number of rabbits currently in the population*. Do you think that many mountain lions go to bed hungry? Explain.

(b) According to equation (1), each increase of the rabbit population by 10 rabbits allows 4 more mountain lions to survive to the next generation. Does this suggest a food surplus or a food deficiency for the mountain lions? Explain. Is this consistent with the answer to (a)?

Remark: The reader might wonder why we discuss a "flawed" model. One answer is that all models are to some extent flawed. In real world problems, there may be too many variables, or **unknown** variables or **unknown**

relationships. A model builder must make choices of which factors are significant and which are not and how they relate to each other. A model builder also must be able to analyze the model produced and be aware that all results it produces are suspect until the model has proven its value. The above model is valuable as an example of this kind of analysis. Actually, the above model is similar to one which does produce credible results.

Specifically, we can produce a more accurate model by replacing equations (1) and (2) with equations of the form

$$R_{n+1} = (1.1)R_n - aR_n M_n$$
$$M_{n+1} = bR_n M_n + (0.7)M_n$$

where a and b are constants to be determined experimentally. The first equation says that the number of rabbits eaten is proportional to *both* the number of rabbits and the number of mountain lions. Thus, if there are few rabbits, then few will be eaten since they are hard to find.

The second equation above says that additional rabbits in the population increases the *percentage* of mountain lions which survive. Equation (1), on the other hand, essentially allows us to change rabbits into mountain lions! The reason that we did not analyze the more accurate model is that it is a non-linear model and this is a *linear* algebra text. The graphical technique we used could, however, have been used on the non-linear model as well.

On Line
2.1 Test for Independence

We will need some functions from the *MatrixManipulation* package.

```
In[1]:= <<LinearAlgebra`MatrixManipulation`
```

■ Discussion

■ Standard Rectangular form for Matrices

Here is a matrix *A*.

```
In[2]:= Clear[A]
        A = {{1, 2}, {3, 4}}
Out[3]= {{1, 2}, {3, 4}}
```

We can display this matrix in the standard form by using *MatrixForm*.

```
In[4]:= MatrixForm[A]
Out[4]//MatrixForm=
        ( 1  2 )
        ( 3  4 )
```

Evaluating the following expression will instruct *Mathematica* to automatically display matrices in the standard rectangular form. This lets us avoid typing *MatrixForm* every time we want to see our matrices in rectangular form rather than as a list of lists. (*$PrePrint* is a global variable whose value is applied to every expression before it is printed. Here we are indicating that if the expression is a matrix then it should be printed in rectangular form and if not then the expression should be printed as usual.)

```
In[5]:= $PrePrint = If[MatrixQ[#], MatrixForm[#], #] &
Out[5]= If[MatrixQ[#1], #1, #1]&
```

2.1 Mathematica On Line.nb

After evaluating this expression we see A in the standard form.

In[6]:= **A**

Out[6]= $\begin{pmatrix} 1 & 2 \\ 3 & 4 \end{pmatrix}$

If we don't want our matrices in rectangular form, (for example, if they are too large for the screen to display clearly) we can evaluate the following expression.

In[7]:= **$PrePrint = .**

This lets us see A as a list of lists again.

In[8]:= **A**

Out[8]= {{1, 2}, {3, 4}}

For now, we want the standard form so we'll turn it on again.

In[9]:= **$PrePrint = If[MatrixQ[#], MatrixForm[#], #] &**

Out[9]= If[MatrixQ[#1], #1, #1]&

■ Column Dependence

Here is the matrix A from Section 2.1, Example 4.

In[10]:= **Clear[A]**
          A = {{1, 2, -1, 3}, {2, 2, -4, 4}, {1, 3, 0, 4}}

Out[11]= $\begin{pmatrix} 1 & 2 & -1 & 3 \\ 2 & 2 & -4 & 4 \\ 1 & 3 & 0 & 4 \end{pmatrix}$

We can find the row reduced form for A.

In[12]:= **RowReduce[A]**

Out[12]= $\begin{pmatrix} 1 & 0 & -3 & 1 \\ 0 & 1 & 1 & 1 \\ 0 & 0 & 0 & 0 \end{pmatrix}$

From the row reduced form, we know that the third and fourth columns are linear combinations of the first two columns. In the row reduced matrix it is clear that

$$column_3 = -3\, column_1 + 1\, column_2$$

and that

$$column_4 = 1\, column_1 + 1\, column_2.$$

The row reduction process does not affect the dependence of the columns so we have the same dependence for the original columns of A. We can verify this as follows. First we extract the columns of A.

In[13]:= **Clear[col1, col2, col3, col4]**

In[14]:= **col1 = TakeColumns[A, {1}]**

Out[14]= $\begin{pmatrix} 1 \\ 2 \\ 1 \end{pmatrix}$

In[15]:= **col2 = TakeColumns[A, {2}]**

Out[15]= $\begin{pmatrix} 2 \\ 2 \\ 3 \end{pmatrix}$

In[16]:= **col3 = TakeColumns[A, {3}]**

Out[16]= $\begin{pmatrix} -1 \\ -4 \\ 0 \end{pmatrix}$

In[17]:= **col4 = TakeColumns[A, {4}]**

Out[17]= $\begin{pmatrix} 3 \\ 4 \\ 4 \end{pmatrix}$

Now we can compare the linear combinations with the third and fourth columns.

2.1 Mathematica On Line.nb

$In[18]:=$ **-3 col1 + 1 col2**

$Out[18]= \begin{pmatrix} -1 \\ -4 \\ 0 \end{pmatrix}$

$In[19]:=$ **1 col1 + 1 col2**

$Out[19]= \begin{pmatrix} 3 \\ 4 \\ 4 \end{pmatrix}$

We see that these linear combinations do, in fact, give us the third and fourth columns.

■ Exercises

1. Let A be the following matrix.

$In[20]:=$ **Clear[A]**
A = {{1, 2, -3}, {4, 5, -1}, {3, 2, 1}, {1, 1, 1}}

$Out[21]= \begin{pmatrix} 1 & 2 & -3 \\ 4 & 5 & -1 \\ 3 & 2 & 1 \\ 1 & 1 & 1 \end{pmatrix}$

Use *RowReduce* to find the row reduced form of A. How can you tell just from this reduced form that the columns of A are independent? Relate your answer to Theorem 1.

2. Let A be a matrix with more rows than columns. State a general rule for using *RowReduce* to decide whether or not the columns of A are independent. Demonstrate your condition by (a) producing a 5 by 4 matrix A with no nonzero coefficients which has dependent columns and computing *RowReduce[A]* and (b) producing a 5 by 4 matrix A with no non-zero coefficients which has dependent columns and computing *RowReduce[A]*. Prove your condition using Theorem 1.

3. Let A be the following matrix.

```
In[22]:= Clear[A]
        A = {{-1, 2, 6, -8, -14, 3},
             {2, 4, 1, -8, 5, -1},
             {-3, 1, 4, -9, -10, 0},
             {3, -2, -1, 12, -1, 4},
             {5, 7, 11, -11, -19, 9}}
```

$$Out[23]= \begin{pmatrix} -1 & 2 & 6 & -8 & -14 & 3 \\ 2 & 4 & 1 & -8 & 5 & -1 \\ -3 & 1 & 4 & -9 & -10 & 0 \\ 3 & -2 & -1 & 12 & -1 & 4 \\ 5 & 7 & 11 & -11 & -19 & 9 \end{pmatrix}$$

Use *RowReduce* to find the pivot columns of A. Write them out explicitly as columns. Then express the other columns of A as linear combinations of the pivot columns. (See the example in the Discussion section.) You should discover that the first three columns of A are the pivot columns.

4. We want to create a matrix B which has the same columns as A but in different order. Here is how to do that.

```
In[24]:= Clear[B]
        B = AppendRows[TakeColumns[A, {4}],
            TakeColumns[A, {1}], TakeColumns[A, {2}],
            TakeColumns[A, {3}], TakeColumns[A, {5}]]
```

$$Out[25]= \begin{pmatrix} -8 & -1 & 2 & 6 & -14 \\ -8 & 2 & 4 & 1 & 5 \\ -9 & -3 & 1 & 4 & -10 \\ 12 & 3 & -2 & -1 & -1 \\ -11 & 5 & 7 & 11 & -19 \end{pmatrix}$$

Find the pivot columns of B by using *RowReduce*. Do you obtain a different set of pivot columns? Use your answer to express the other columns as linear combinations of the pivot columns. Could you have derived these expressions from those in the previous exercise? If so, how?

5. Find a matrix B_2 whose columns are just those of A listed in a different order, such that the column of B_2 which equals the fifth column of A and the column of

B_2 which equals the first column of A are both pivot columns. Is it possible to find such a B_2 where the second column of A is a pivot column as well? If so, find an example. If not, explain why it is not possible.

On Line
2.2 Dimension

We want to view our matrices in standard form.

In[1]:= **$PrePrint = If[MatrixQ[#], MatrixForm[#], #] &**

Out[1]= If[MatrixQ[#1], #1, #1]&

■ Discussion

Mathematica does not contain a "rank" function which returns the rank of a matrix A. However, we can define one ourselves. The rank of A is the number of nonzero rows in the reduced echelon form of A. Thus, the rank is equal to the number of basic variables. We can compute the number of basic variables by subtracting the number of free variables, *Length[NullSpace[A]]*, from the total number of variables. The total number of variables is the number of columns of A or the number of rows of the transpose of A, *Length[Transpose[A]]*.

In[2]:= **Clear[rank, A]**
         rank[A_] := Length[Transpose[A]] -
                      Length[NullSpace[A]]

■ Exercises.

1. Here is a matrix A.

```
In[4]:= Clear[A]
        A = {{-1, 2, 6, -8, -14, 3},
             {2, 4, 1, -8, 5, -1},
             {-3, 1, 4, -9, -10, 0},
             {3, -2, -1, 12, -1, 4},
             {5, 7, 11, -11, -19, 9}}
```

$$Out[5]= \begin{pmatrix} -1 & 2 & 6 & -8 & -14 & 3 \\ 2 & 4 & 1 & -8 & 5 & -1 \\ -3 & 1 & 4 & -9 & -10 & 0 \\ 3 & -2 & -1 & 12 & -1 & 4 \\ 5 & 7 & 11 & -11 & -19 & 9 \end{pmatrix}$$

Use the function *rank* defined above to find the rank of A. *Using only the value of the rank*, explain why statements (a) and (b) are true. Then do the rest of the parts.

(a) The reduced form for the augmented matrix for the system $A \cdot x = 0$ has three free variables.

(b) The nullspace of A has dimension at most 3. (Hint: How many spanning vectors are there in the general solution to $A \cdot x = 0$?)

(c) Show that each of the vectors below satisfy $A \cdot x = 0$.

```
In[6]:= Clear[x1, x2, x3]
        x1 = {-5, 13, -10, 2, -3, 1}
Out[7]= {-5, 13, -10, 2, -3, 1}

In[8]:= x2 = {3, -6, 11, 1, 2, -5}
Out[8]= {3, -6, 11, 1, 2, -5}

In[9]:= x3 = {-4, 7, 9, 5, 1, -6}
Out[9]= {-4, 7, 9, 5, 1, -6}
```

(d) Prove that x_1, x_2, and x_3 are linearly independent by computing the rank of the matrix X whose rows are x_1, x_2, and x_3.

In[10]:= X = {x1, x2, x3}

$$Out[10] = \begin{pmatrix} -5 & 13 & -10 & 2 & -3 & 1 \\ 3 & -6 & 11 & 1 & 2 & -5 \\ -4 & 7 & 9 & 5 & 1 & -6 \end{pmatrix}$$

(e) How does it follow that the dimension of the nullspace of A is 3? How does it follow that x_1, x_2, and x_3 constitute a basis for the nullspace?

(f) Use *NullSpace* to find a basis for the nullspace of A. Express each of these vectors as a linear combination of the vectors x_1, x_2, and x_3.

In[11]:= {y1, y2, y3} = NullSpace[A]

$$Out[11] = \begin{pmatrix} -1 & 1 & -1 & 0 & 0 & 1 \\ 0 & -2 & 3 & 0 & 1 & 0 \\ -2 & 3 & 0 & 1 & 0 & 0 \end{pmatrix}$$

Remark: An efficient way of doing this is as follows. Begin by defining the matrix B whose first three columns are the vectors x_1, x_2, and x_3 and whose last three columns are the basis vectors, y_1, y_2, y_3, for the null space of A.

In[12]:= B = Transpose[{x1, x2, x3, y1, y2, y3}]

$$Out[12] = \begin{pmatrix} -5 & 3 & -4 & -1 & 0 & -2 \\ 13 & -6 & 7 & 1 & -2 & 3 \\ -10 & 11 & 9 & -1 & 3 & 0 \\ 2 & 1 & 5 & 0 & 0 & 1 \\ -3 & 2 & 1 & 0 & 1 & 0 \\ 1 & -5 & -6 & 1 & 0 & 0 \end{pmatrix}$$

The matrix B should have rank 3. (Why?) If you reduce B, the first three columns will be the pivot columns. According to Theorem 1 in Section 2.1, the last three columns will then be linear combinations of the first three. In fact, the technique of Section 2.1 *Mathematica* On Line shows how to find the explicit expressions for the y_i in terms of the x_i.

(g) In part (f), what made us so sure that the first three columns would be the pivot columns? Why couldn't, for example, the pivot columns be columns 1, 3

2.2 Mathematica On Line.nb

and 4? (Hint: Think about what this would say about the reduced form of the matrix whose columns are x_1, x_2, and x_3.)

(h) Express each of x_1, x_2, and x_3 as a linear combination of y_1, y_2, and y_3. This demonstrates that x_1, x_2, and x_3 (the spanning vectors from the solution of $A \cdot x = 0$) and y_1, y_2, and y_3 (the spanning vectors from using *NullSpace*) both span the nullspace.

(i) Find, using *RowReduce* on an augmented matrix, a vector t which solves the equation $A \cdot x = \{6, 1, 4, 1, 11\}$.

(j) Let $z = t + Random[\,] \, x_1 + Random[\,] \, x_2 + Random[\,] \, x_3$ where t is as in (k). Compute $A \cdot z$. Explain why you get what you get what you get. Find constants u, v, and w such that $z = t + u \, y_1 + v \, y_2 + w \, y_3$. What theorem does this demonstrate?

On Line
2.2.1 Differential Equations

■ **Discussion**

Imagine that we have a box of mass 1g on a frictionless table and attached to a spring on the left. Initially, the spring is un-stretched. We pull the box 3 cm to the right and let it go. We expect that it will oscillate left and right across the table. In this exercise set, we would like to quantify this.

The physical law which we use is

$$F = ma$$

where F is the force on the box, m is its mass (so, in our case, $m = 1$) and a is its acceleration. Explicitly, we let $y(t)$ denote the displacement of the box away from its rest position at time t. (Displacements to the right are considered to be positive.) Then, acceleration at time t is $y''(t)$.

2.2.1 Mathematica On Line.nb

The only force on the box which acts in the direction of the motion is that exerted by the spring. We shall assume that this force is proportional to the stretching. Specifically, we shall assume that this force is approximately 0.25 dynes per centimeter of stretching. This force will, of course, also be in the opposite direction from the stretching. Thus, $F = ma$ translates to

$$-0.25y = y''$$

which is equivalent with

$$y'' + 0.25\, y = 0. \qquad (7)$$

This is a differential equation similar to the ones studied in this section.

Mathematica is able to symbolically solve this differential equation with the function *DSolve*. We must specify the differential equation, the function, and the independent variable. (Note the use of double equal signs, "==", to indicate that we are working with an equation and not a variable assignment.)

```
In[1]:= Clear[y,t]
        DSolve[{y''[t] + 0.25y[t] == 0},y[t],t]

Out[2]= (y[t] → C[2] Cos[0.500000000000000 t] -
              1.00000000000000 C[1] Sin[0.500000000000000 t])
```

In our case we also have two initial conditions, $y[0] = 3$ and $y'[0] = 0$. To solve a differential equation with initial conditions we write the equation and the conditions in a list and use *DSolve*. (Again, we use double equal signs, "==".)

```
In[3]:= Clear[y, t, solution]
        solution = DSolve[{y''[t] + 0.25y[t] == 0,
                   y'[0] == 0, y[0] == 3}, y[t], t]

Out[4]= (y[t] → 3.00000000000000 Cos[0.500000000000000 t])
```

Here the solution is given as a replacement for $y[t]$. We can define a function that is the solution by applying the rules generated by *DSolve* to $y[t]$ using the replacement operator.

In[5]:=
```
Clear[y,t]
y[t_] = y[t] /. solution[[1,1]]
```
Out[6]= 3.00000000000000 Cos[0.50000000000000 t]

Now we can plot the solution on the interval from 0 to 40.

In[7]:= `Plot[y[t],{t,0,40}]`

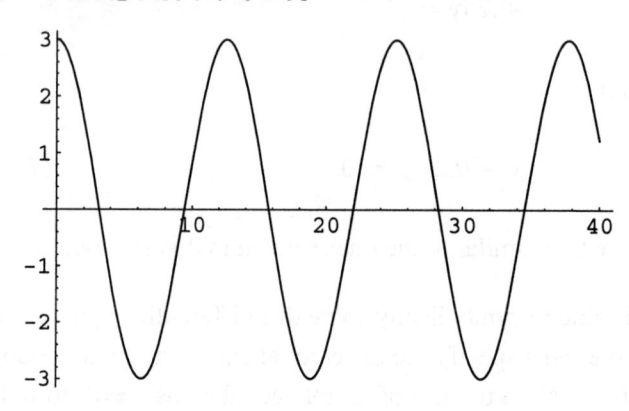

Out[7]= - Graphics -

We can also plot both the function and its derivative on the same interval.

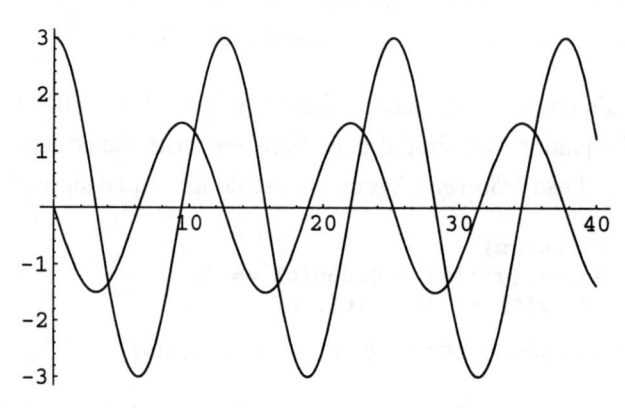

In[8]:= `Plot[{y[t], y'[t]}, {t, 0, 40}]`

Out[8]= - Graphics -

2.2.1 Mathematica On Line.nb

■ Exercises.

1. The period of a periodic motion is the amount of time before the motion repeats itself. Use the graph of y to estimate the period of the motion. Would you consider this "slow" or "fast" oscillation? (The time is in seconds.)

2. Let's take another look at the solution to our differential equation.

In[9]:= **y[t]**

Out[9]= 3.00000000000000 Cos[0.50000000000000 t]

From the formula for y, find the exact value for the period of the motion you estimated in the previous exercise. Comare your results.

3. The period of the oscillations is determined by the "stiffness" of the spring. (A stiff spring is one which takes a large force to stretch it.) How do you guess stiffness should relate to the period of the motion: should stiff springs oscillate faster or slower? Test your guess by graphing the solution curve for a stiff spring and a non-stiff spring. Note: This will require that you change the coefficient 0.25 in equation (7). How should you change it to model a stiffer spring?

4. Prove mathematically that your guess in the last exercise is really correct. For this, you should compute the general solution to the equation $y'' = ky$ and then find a formula for the period.

5. Imagine that our box is sitting near our stereo which is generating a tone which is causing the box to vibrate. We model this as applying an external force $F(t) = \varepsilon \sin(\omega t)$ where ω is determined by the pitch of the tone and ε is determined by the volume of the tone. This changes formula (7) to

$$-0.25y + \varepsilon \sin(\omega t) = y'' \qquad (8)$$

Assume initially that $\varepsilon = 0.4$ and $\omega = 0.3$. Plot the solution to this equation over the interval $0 \leq t \leq 100$. Describe the motion of the box. What is its maximum displacement?

6. Repeat the previous exercise with $\varepsilon = 0.4$ and $\omega = 0.5$. Plot the solution over a large enough interval to allow you to guess the maximum value of the displacement.

7. Surely, the behavior described by the graph in the previous exercise would not happen with a real box. What factors would limit the actual displacement of the box?

Remark: Exercises 5 and 6 demonstrate the concept of resonance: if we vibrate the spring with just the right tone, the displacements can become very large. The same principle applies to structures such as bridges. The most famous example of this occurred on November 7, 1940 when the Tacoma Narrows bridge at Puget Sound self-destructed in a high wind.

On Line
2.3 Applications to systems

We want to view our matrices in standard form.

In[1]:= **$PrePrint = If[MatrixQ[#], MatrixForm[#], #] &**

Out[1]= If[MatrixQ[#1], #1, #1]&

We will need the *MatrixManipulation* package.

In[2]:= **<<LinearAlgebra`MatrixManipulation`**

We will also need our function *rank*.

In[3]:= **Clear[rank, A]**
  **rank[A_] := Length[Transpose[A]] -
       Length[NullSpace[A]]**

■ Discussion

We can create a matrix with random entries with the function *Table*. Here is a 2 by 4 example.

In[5]:= **Clear[A]**
  A = Table[Random[], {i, 1, 2}, {j, 1, 4}]

$$Out[6]= \begin{pmatrix} 0.194309 & 0.0847198 & 0.773246 & 0.144086 \\ 0.212568 & 0.71273 & 0.684031 & 0.833584 \end{pmatrix}$$

We can row reduce this matrix.

In[7]:= **reducedA = RowReduce[A]**

$$Out[7]= \begin{pmatrix} 1 & 0. & 4.0933 & 0.26621 \\ 0 & 1 & -0.261071 & 1.09017 \end{pmatrix}$$

From this we see that the third and fourth columns of *A* are dependent upon the first two columns of *A*. We can use the entries of the matrix *reducedA* to demonstrate that dependence. First we need to extract the columns of *A*.

```
In[8]:= col1 = TakeColumns[A,{1}];
        col2 = TakeColumns[A,{2}];
        col3 = TakeColumns[A,{3}];
        col4 = TakeColumns[A,{4}];
```

The third column of *reducedA* gives the dependency coefficients. We evaluate the following expression and compare the results with the third column of *A*.

```
In[12]:= reducedA[[1,3]] col1 + reducedA[[2,3]] col2
```

$$Out[12]= \begin{pmatrix} 0.773246 \\ 0.684031 \end{pmatrix}$$

```
In[13]:=
         col3
```

$$Out[13]= \begin{pmatrix} 0.773246 \\ 0.684031 \end{pmatrix}$$

Similarly, the fourth column can be written as a linear combination of the first two columns.

```
In[14]:= reducedA[[1,4]] col1 + reducedA[[2,4]] col2
```

$$Out[14]= \begin{pmatrix} 0.144086 \\ 0.833584 \end{pmatrix}$$

```
In[15]:=
         col4
```

$$Out[15]= \begin{pmatrix} 0.144086 \\ 0.833584 \end{pmatrix}$$

The reason we are able to find the coefficients from *reducedA* is that in obtaining the reduced form of *A* we do not change any of the dependencies among the columns. In *reducedA* the entries in the third column clearly indicate what multiples of the first and second columns are required to add up to the third column. The same dependency will exist in the original matrix *A*. (Similar results are also clear for the fourth column.)

2.3 Mathematica On Line.nb

■ **Exercises.**

1. Construct a random 3 by 5 matrix *M*.

$In[16]:=$ **Clear[M]**
  M = Table[Random[], {i, 1, 3}, {j, 1, 5}]

$Out[17]=$ $\begin{pmatrix} 0.0935631 & 0.869915 & 0.751301 & 0.865774 & 0.484394 \\ 0.953325 & 0.0337416 & 0.591538 & 0.119097 & 0.223765 \\ 0.891302 & 0.306122 & 0.801779 & 0.751898 & 0.621601 \end{pmatrix}$

What do you expect the rank of *M* to be? Check your guess using the function *rank*. Is it conceivable that the rank could have turned out otherwise? Why is it unlikely?

2. We will now extend the matrix *M* from the previous problem to a 5 by 5 matrix M_2 where the fourth and fifth rows are random linear combinations of the three rows of *M*. We use curly brackets when defining the new rows because the function *AppendColumns* requires matrices and not vectors for its entries. Thus, *row4* and *row5* will be 1 by 5 matrices and not vectors from $\mathbf{R}^5$.

$In[18]:=$ **row4 = {Random[] M[[1]] + Random[] M[[2]] +
        Random[] M[[3]]};
  row5 = {Random[] M[[1]] + Random[] M[[2]] +
        Random[] M[[3]]};
  M2 = AppendColumns[M, row4, row5]**

$Out[20]=$ $\begin{pmatrix} 0.0935631 & 0.869915 & 0.751301 & 0.865774 & 0.484394 \\ 0.953325 & 0.0337416 & 0.591538 & 0.119097 & 0.223765 \\ 0.891302 & 0.306122 & 0.801779 & 0.751898 & 0.621601 \\ 1.17397 & 0.226532 & 0.895823 & 0.575787 & 0.55165 \\ 1.24929 & 0.887835 & 1.46152 & 1.13359 & 0.848396 \end{pmatrix}$

Note that M_2 is a 5 by 5 matrix. Use the function *rank* to determine the rank of M_2?

Check this result by using *RowReduce*.

Are your results in agreement? If not, explain why they are inconsistent. You may be experiencing error because of roundoff problems in the computation of the row reduced form. The option *ZeroTest* will help deal with the roundoff errors. As used here, all values less than 10^{-10} are automatically set equal to 0.

In[21]:= **RowReduce[M2, ZeroTest -> (Chop[#] == 0 &)]**

$$Out[21]= \begin{pmatrix} 1 & 0 & 0 & -7.44405 & -5.1989 \\ 0 & 1 & 0 & -9.19189 & -6.7809 \\ 0 & 0 & 1 & 12.7225 & 9.14364 \\ 0 & 0 & 0 & 0 & 0 \\ 0 & 0 & 0 & 0 & 0 \end{pmatrix}$$

What is the maximal number of linearly independent *columns* in M_2?

3. For the 5 by 5 matrix M_2 above, find a set of columns of M_2 which forms a basis for the column space. Express the other columns of M as linear combinations of these columns.

4. For M_2 as above, find a basis for the column space of M_2 by reducing the transpose of M_2 and then using the Non-Zero Rows Theorem. You will need to use the *ZeroTest* option. We'll start by finding the transpose of M_2.

In[22]:= **transposeM2 = Transpose[M2]**

$$Out[22]= \begin{pmatrix} 0.0935631 & 0.953325 & 0.891302 & 1.17397 & 1.24929 \\ 0.869915 & 0.0337416 & 0.306122 & 0.226532 & 0.887835 \\ 0.751301 & 0.591538 & 0.801779 & 0.895823 & 1.46152 \\ 0.865774 & 0.119097 & 0.751898 & 0.575787 & 1.13359 \\ 0.484394 & 0.223765 & 0.621601 & 0.55165 & 0.848396 \end{pmatrix}$$

Try to express each of the basis columns you found in Exercise 3 as linear combinations of these columns.

5. According to the comments in the text, there is an inverse relationship between the dimension of the nullspace of a matrix and its rank. Demonstrate this by creating (as in Exercise 1 above) 4 random 4×4 matrices with rank 1, 2, 3, and 4 respectively. For each of your matrices use the function *NullSpace* to find a basis for the nullspace.

On Line
3.1 Linear Transformations

We want to view our matrices in standard form.

```
In[1]:= $PrePrint = If[MatrixQ[#], MatrixForm[#], #] &

Out[1]= If[MatrixQ#1], #1, #1]&
```

■ Discussion

The following definition gives parametric equations in three intervals that will produce the letter "R" when they are plotted. The conditions for the constrained definitions follow the symbol "/;" and we use ":=" instead of "=" because we want to delay the definition until the function is called.

```
In[2]:= Clear[x, y, t]
        x[t_] := 0                        /; 0 <= t < 1
        y[t_] := t                        /; 0 <= t < 1

        x[t_] := Sin[Pi (t - 1)]          /; 1 <= t < 2
        y[t_] := 3/4 + Cos[Pi (t - 1)]/4  /; 1 <= t < 2

        x[t_] := t - 2                    /; 2 <= t <= 3
        y[t_] := (3 - t)/2                /; 2 <= t <= 3
```

Now that we have the definitions for the parametric equations we can plot them using *ParametricPlot*. The *PlotStyle* option specifies that the scalings on the *x*- and *y*-axes are the same.

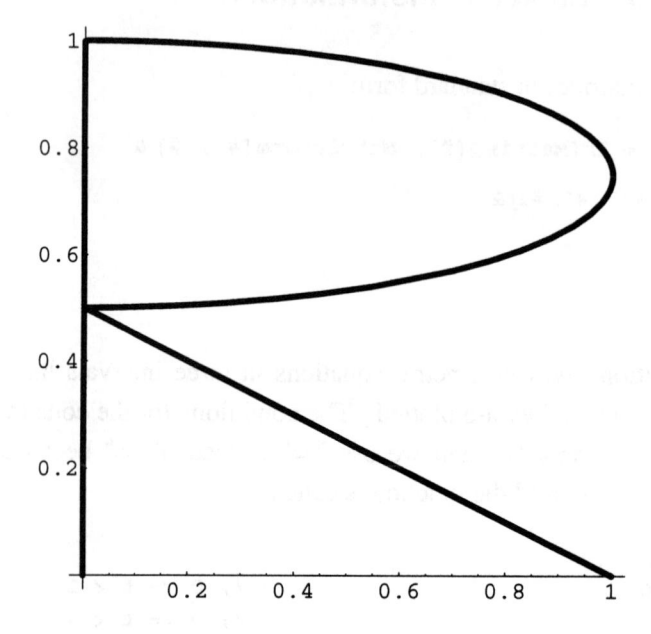

```
In[9]:= ParametricPlot[{x[t], y[t]}, {t, 0, 3},
          PlotStyle -> {Thickness[0.01]},
          AspectRatio -> Automatic]
```

Out[9]= - Graphics

We referred to the transformation defined by the matrix in Example 1 of Section 3.1 of your test as a "shear along the *x*-axis." Here is the matrix from that example.

```
In[10]:= M = {{1, 0}, {1, 1}}
```

$Out[10]= \begin{pmatrix} 1 & 0 \\ 1 & 1 \end{pmatrix}$

To see the effect of this shearing on our "R" we can multiply the vector of parametric equations by the matrix *M* and plot the result.

3.1 Mathematica On Line.nb

$In[11]:=$ **ParametricPlot[M . {x[t], y[t]}, {t, 0, 3},**
   PlotStyle -> {Thickness[0.02]},
   AspectRatio -> Automatic]

ParametricPlot::ppcom : Function M.{x[t], y[t]} cannot be compiled; plotting will proceed with the uncompiled function.

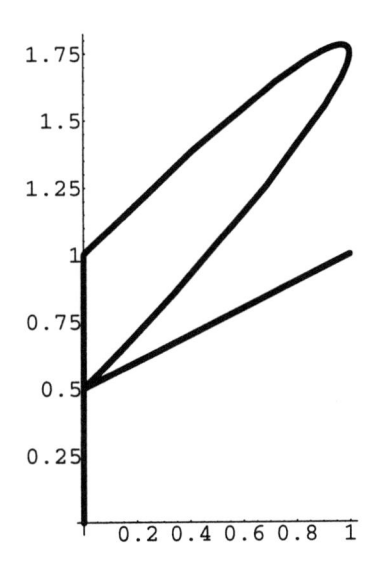

$Out[11]=$ - Graphics

If we want to shift the "R" in the x or y directions we can just add appropriate constants to the parametric equations we graph. Here is an example where we plot our original "R" and a copy that has been shifted 1 unit to the right and 2 units vertically.

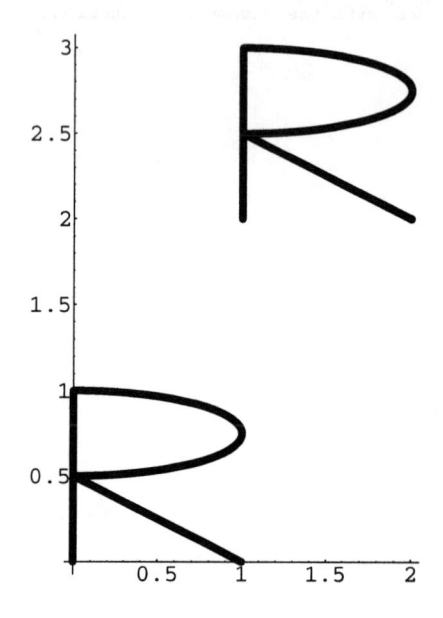

```
In[12]:= ParametricPlot[{{x[t], y[t]},
         {x[t] + 1, y[t] + 2}}, {t, 0, 3},
         PlotStyle -> {Thickness[0.02]},
         AspectRatio -> Automatic]
```

Out[12]= - Graphics

■ Exercises

1. Use the rotation matrix to plot the image of "R" under rotation by 20 degrees. (Note: *Mathematica* works in radians so you will need to convert from degrees to radians.)

2. Let S be the transformation of R^2 into itself defined by stipulating that $S(x)$ is the result of shifting x one unit to the right. Show graphically that S is not linear. Specifically, use our "R" to show that $S(2x) \neq 2S(x)$.

3. For each item below, find a matrix M for which multiplication by M would give the specified linear transformation. Plot the transformed "R."

(a) flip the "R" upside down;

(b) flip the "R" left-to-right;

(c) rotate the "r" by 30 degrees;

(d) shear the "R" along the y-axis.

4. Plot the effects on our "R" of the following transformations:

(a) A shear along the x-axis followed by a rotation by 20 degrees;

(b) A rotation of 20 degrees followed by a shear along the x-axis;

(c) A shear along the x-axis followed by a shear along the y-axis;

(d) A shear along the y-axis followed by a shear along the x-axis.

On Line
3.2 Multiplication

We want to view our matrices in standard form.

```
In[1]:= $PrePrint = If[MatrixQ[#], MatrixForm[#], #] &
Out[1]= If[MatrixQ#1], #1, #1]&
```

■ Discussion

We will be using matrix multiplication to move graphics objects in three dimensions. First, we create a three dimensional letter "R" in the *xy*-plane by defining the following parametric equations.

```
In[2]:= Clear[x, y, t]
        x[t_] := 0                          /; 0 <= t < 1
        y[t_] := t                          /; 0 <= t < 1
        z[t_] := 0

        x[t_] := Sin[Pi (t-1)]              /; 1 <= t < 2
        y[t_] := 3/4 + Cos[Pi (t-1)]/4      /; 1 <= t < 2
        z[t_] := 0

        x[t_] := (t-2)                      /; 2 <=t <= 3
        y[t_] := (3-t)/2                    /; 2 <= t <= 3
        z[t_] := 0
```

We can display the "R" by using *ParametricPlot3D*. The function *Evaluate* compiles our functions and makes the computations quicker. We plot our original "R" in red so that we can distinguish it from copies which have been moved. We label the axes with the option *AxesLabel*. Finally, we label this plot *r* so that we can show it again along with copies that have been moved.

3.2 Mathematica On Line.nb   125

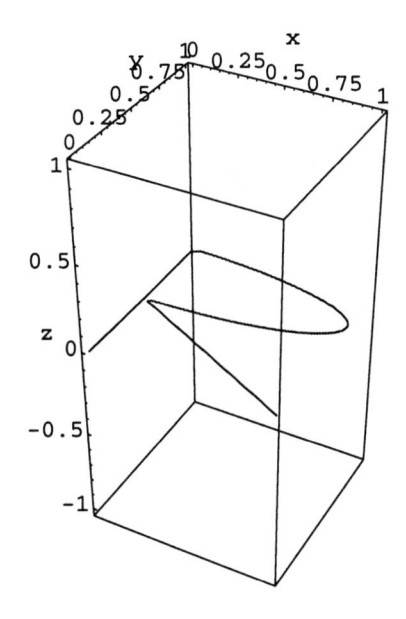

```
In[12]:= Clear[r]
         r = ParametricPlot3D[Evaluate[{x[t], y[t], z[t],
           RGBColor[1,0,0]}], {t, 0, 3},
           AxesLabel -> {"x", "y", "z"}]
```

Out[13]= - Graphics3D -

We can rotate this "R" by Pi/6 radians about the *x*-axis by means of the following matrix multiplication. First, we define the matrix.

```
In[14]:= Clear[m]
         m = {{1,0,0},{0,Cos[Pi/6],-Sin[Pi/6]},
           {0,Sin[Pi/6],Cos[Pi/6]}}
```

$$Out[15]= \begin{pmatrix} 1 & 0 & 0 \\ 0 & \frac{\sqrt{3}}{2} & -\frac{1}{2} \\ 0 & \frac{1}{2} & \frac{\sqrt{3}}{2} \end{pmatrix}$$

We define new parametric equations given by multiplication by *m*.

```
In[16]:= Clear[x2, y2, z2, t]
         {x2[t_], y2[t_], z2[t_]} =
            m . {x[t], y[t], z[t]}
```

$Out[17]= \left\{x[t], \frac{1}{2}\sqrt{3}\ y[t], \frac{y[t]}{2}\right\}$

Finally, we plot the new parametric equations.

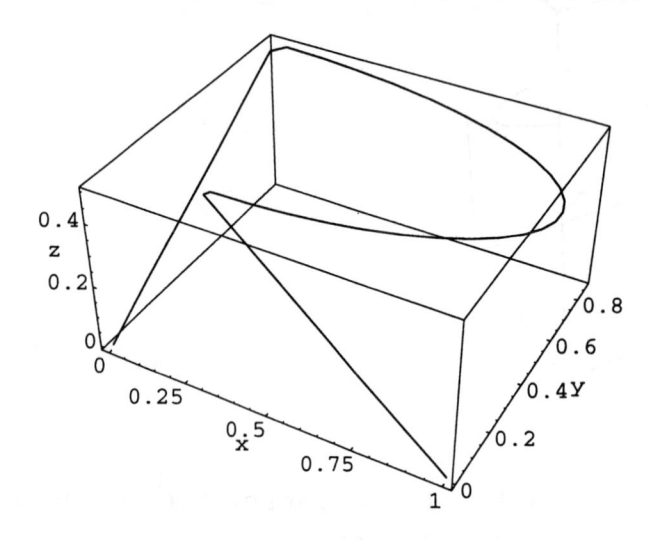

```
In[18]:= Clear[r2]
         r2 = ParametricPlot3D[Evaluate[{x2[t],y2[t],z2[t]}],
            {t,0,3}, AxesLabel -> {"x", "y", "z"}]
```

Out[19]= - Graphics3D -

We can show both "R"s at the same time.

3.2 Mathematica On Line.nb

In[20]:= **Show[r,r2]**

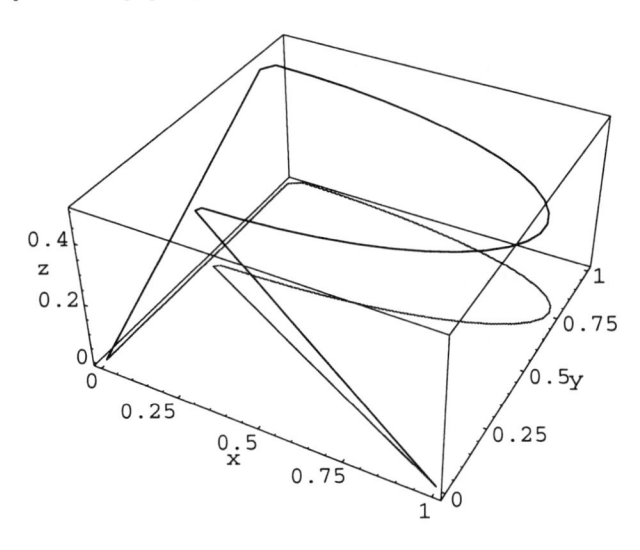

Out[20]= - Graphics3D -

■ Exercises

1. Rotate "R" by 30 degrees about the *x*-axis and then rotate this image by 20 degrees about the *z*-axis. Plot both images.

2. Find a single matrix which transforms the "R" into the final image from the last exercise. Plot the image of "R" under this transformation.

3. How do you suppose the image would appear if you were to transform the result of the previous exercise by a rank 2 matrix transformation? Create a random rank 2 matrix and test your guess. Show both your beginning image and final image.

4. How do you suppose the image would appear if you were to transform our original "R" by a rank 1 matrix transformation? Create a random rank 1 matrix and test your guess.

On Line
3.3 Image

We will need the Matrix Manipulation package.

In[1]:= `<<LinearAlgebra`MatrixManipulation` `

We want our matrices to appear in standard form.

In[2]:= `$PrePrint = If[MatrixQ[#], MatrixForm[#], #] &`

Out[2]= `If[MatrixQ#1], #1, #1]&`

■ Discussion

Here is a 2 by 3 matrix with rank 1.

In[3]:= `Clear[M]`
`M = {{1, 2, 3}, {2, 4, 6}}`

Out[4]= $\begin{pmatrix} 1 & 2 & 3 \\ 2 & 4 & 6 \end{pmatrix}$

Here is a list of the images of 100 random points in $\mathbf{R}^3$ under multiplication by M.

In[5]:= `images = Table[M . {Random[], Random[], Random[]},`
`{i, 1, 100}];`

We can plot these points as follows:

In[6]:= **ListPlot[images]**

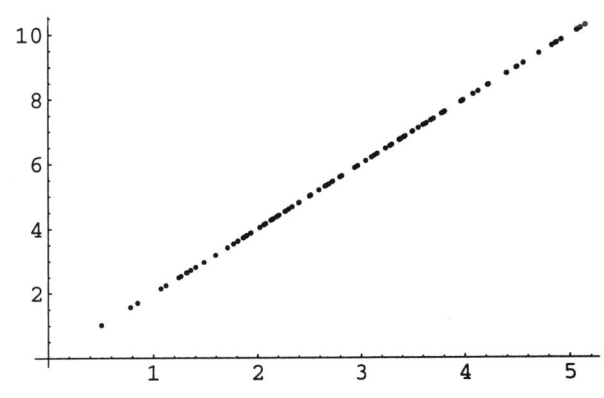

Out[6]= - Graphics

We can extract the first column of *M* in the following way.

In[7]:= **column = Transpose[M][[1]]**

> General::spell1 : Possible spelling error: new symbol
>   name "column" is similar to existing symbol "Column".

Out[7]= {1, 2}

We can plot 100 elements in the span of this column as follows:

In[8]:= **Clear[span]**
       span = Table[Random[] column, {k, 1, 100}];

Finally, we can graph these elements in the column space of *M*.

In[10]:= **ListPlot[span]**

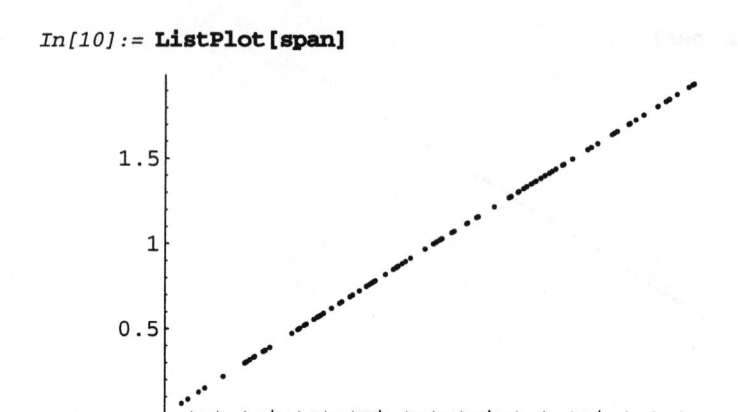

Out[10]= - Graphics

We can find the nullspace of *M* using *NullSpace*.

In[11]:= **Clear[basis]**
         basis = NullSpace[M]

$Out[12]= \begin{pmatrix} -3 & 0 & 1 \\ -2 & 1 & 0 \end{pmatrix}$

The basis vectors for the nullspace are the rows of this matrix. We can extract the rows as follows:

In[13]:= **vector1 = basis[[1]]**

Out[13]= {-3, 0, 1}

In[14]:= **vector2 = basis[[2]]**

Out[14]= {-2, 1, 0}

Here is a plot of 100 random vectors in the null space of *M*.

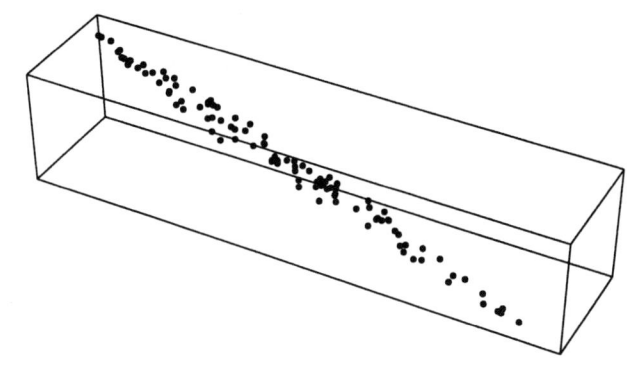

```
In[15]:= Show[Table[
           Graphics3D[Point[Random[] vector1 + Random[] vector2]],
           {n, 1, 100}]]
```

[3D plot of 100 random points forming a line segment within a rectangular box]

```
Out[15]= - Graphics3D
```

■ Exercises

1. Create a "random" 2 by 3 matrix M with rank 1. (See Exercise 2 in On Line Section 2.3 for information on creating random matrices with specific ranks.) What should the dimension of the image of the transformation defined by M be? Verify this by plotting the image under multiplication by M of 100 random points in R^3. Question: Why are you getting only a line segment for the image rather than a whole line? How can you get more of the line?

2. In the previous exercise, the image should be the span of any non-zero column of M. Demonstrate ths by choosing a column of M and plotting 100 random points in its span. Try to plot a large part of the span, not just a small segment.

3. Use your information from the previous exercises to (a) find a specific vector in R^2 such that the equation $M \cdot x = b$ is not solvable and (b) find a specific vector c in R^2 such that the equation $M \cdot x = c$ is solvable. Indicate where these vectors would lie in the graph from Exercise 1 or 2. Verify your answers by applying *RowReduce* to the augmented matrices [M b] and [M c].

4. Plot 100 random elements of the nullspace of M. Explain how your plot relates to the Rank-Nullity Theorem.

5. Create a "random" 3 by 3 matrix M with rank 2. Repeat the last three exercies (suitably modified to take account of the different dimensions) for this M.

On Line
3.4 Inverses

We want our matrices in traditional form.

In[1]:= `$PrePrint = If[MatrixQ[#], MatrixForm[#], #] &`

Out[1]= If[MatrixQ[#1], #1, #1]&

■ Discussion

■ Solving Nonsingular Linear Systems

We can solve systems of the form $A \cdot x = b$ using the *Mathematica* function *LinearSolve*. Here is an example.

In[2]:= `Clear[A,b]`
`A = {{2, 3}, {4, -2}}`
`b = {5, -2}`

Out[3]= $\begin{pmatrix} 2 & 3 \\ 4 & -2 \end{pmatrix}$

Out[4]= {5, -2}

In[5]:= `x = LinearSolve[A, b]`

Out[5]= $\{\frac{1}{4}, \frac{3}{2}\}$

We can check our solution.

In[6]:= `A . x`

Out[6]= {5, -2}

Alternatively, we could find the inverse of *A* and use it to find the solution to the linear system. We'll let *Ainv* be the inverse.

$In[7]:=$ **Clear[Ainv]**
Ainv = Inverse[A]

$Out[8]= \begin{pmatrix} \frac{1}{8} & \frac{3}{16} \\ \frac{1}{4} & -\frac{1}{8} \end{pmatrix}$

$In[9]:=$ **Ainv . b**

$Out[9]= \{\frac{1}{4}, \frac{3}{2}\}$

Compare this with our previous solution x.

$In[10]:=$ **x**

$Out[10]= \{\frac{1}{4}, \frac{3}{2}\}$

■ Vector Norm

Mathematica does not have a function for computing the norm of a vector, but we can build one of our own. Since $\|v\|^2 = v \cdot v$, we can define *norm* as follows:

$In[11]:=$ **Clear[norm, v]**
norm[v_] = Sqrt[v . v]

$Out[12]= \sqrt{v.v}$

Here is an example of how we can use it.

$In[13]:=$ **v = {1, 2, 3}**
norm[v]

$Out[13]= \{1, 2, 3\}$

$Out[14]= \sqrt{14}$

■ Exercises

1. Here is the matrix from Exercise 2(a) of Section 3.4.

3.4 Mathematica On Line.nb

```
In[15]:= Clear[A]
         A = {{1, 0, 3}, {4, 4, 2}, {2, 5, -4}}
```

$$Out[16] = \begin{pmatrix} 1 & 0 & 3 \\ 4 & 4 & 2 \\ 2 & 5 & -4 \end{pmatrix}$$

Here is a vector.

```
In[17]:= b = {2.1, 3.2, -4.4}
Out[17]= {2.1, 3.2, -4.4}
```

Use *LinearSolve* to solve the system $A \cdot x = b$.

2. In most applications of linear algebra, our numerical data comes from measurements which are susceptible to error. Suppose that the vector b in the previous problem was obtained by measuring a vector b_2 whose actual value is given as follows:

```
In[18]:= b2 = {2.1, 3.21, -4.4}
Out[18]= {2.1, 3.21, -4.4}
```

Compute the solution to $A \cdot x_2 = b_2$ using *LinearSolve*.

Which component of the x you computed in the previous exercise has the largest error? (We measure error as the absolute value of the difference between the computed value and the actual value.) Explain why this is to be expected in terms of the magnitude of the components of A^{-1}. Which component of b would you change in order to produce the greatest change in x? Why? Back up your answer with a numerical example. How much error could you tolerate in the measured values of the components of b if each entry of x is to have an error of at most ±0.001?

2. Here is a new matrix and a new vector.

```
In[19]:= Clear[A, b]
        A = {{1, 1/2, 1/3}, {1/2, 1/3, 1/4},
             {1/3, 1/4, 1/5}}
        b = {83, 46, 32}
```

$$Out[20]= \begin{pmatrix} 1 & \frac{1}{2} & \frac{1}{3} \\ \frac{1}{2} & \frac{1}{3} & \frac{1}{4} \\ \frac{1}{3} & \frac{1}{4} & \frac{1}{5} \end{pmatrix}$$

Out[21]= {83, 46, 32}

Use *LinearSolve* to compute the solution to $A \cdot x = b$.

As in the previous exercise, suppose that the vector b was obtained by measuring a vector b_2 whose actual values are given as follows:

```
In[22]:= b2 = {82.9, 46.07, 31.3}
```

Out[22]= {82.9, 46.07, 31.3}

Use *LinearSolve* to compute the solution to $A \cdot x_2 = b_2$.

What is the percentage of error in the least accurate entry of x? What is there about the components of A^{-1} which accounts for the large error? How much error could you tolerate in the measured values of the components of b if each entry of x is to have an error of at most ±0.001?

4. The above exercise demonstrates that the process of solving a system of equations can "magnify" errors in disastrous ways. One quantitative measure of the **inaccuracy** of a calculation is the *ratio of the percentage of error of the final answer to the percentage of error of the input data*. But what do we mean by the percentage of error in a vector (such as x in Exercise 1 above) in which every component might have errors of different magnitudes?

For vectors in R^3, this question has a geometric answer. We think of x and y as representing points in R^3. The distance d between these points is one measure of the error. If $x = \{x_1, x_2, x_3\}$ and $y = \{y_1, y_2, y_3\}$, then

3.4 Mathematica On Line.nb

$$d = norm[x - y].$$

Now if x is the computed answer and y is the actual answer, we define the **percentage of error** to be

$$P = 100\, norm[x - y] / norm[x]$$

a. Let b, b_2, x and x_2 be as in Exercise 2 above. Use the above formula to compute (i) the percentage of error in b, (ii) the percentage of error in x and (iii) the ratio of the percentage of error in x to that in b. This is the inaccuracy of the calculation of x from b.

Question: Assuming that accuracy is desired, do we want this number to be large or small? Explain.

b. Compute the inaccuracy of the computation of x from b in Exercise 3 above.

On Line
3.5 LU-Factorization

We want to view our matrices in standard form.

In[1]:= **$PrePrint = If[MatrixQ[#], MatrixForm[#], #] &**

Out[1]= If[MatrixQ[#1], #1, #1]&

■ Discussion

Here is a matrix A and a vector b.

In[2]:= **Clear[A, b]**
A = {{1, 2, 1}, {1, 3, 4}, {2, 7, 8}}
b = {1, 2, 3}

$$Out[3]= \begin{pmatrix} 1 & 2 & 1 \\ 1 & 3 & 4 \\ 2 & 7 & 8 \end{pmatrix}$$

Out[4]= {1, 2, 3}

We can solve the system $A \cdot x = b$ by using the LU decomposition of the matrix A. The *Mathematica* function *LUDecomponsition* performs the LU decomposition of A and returns a list of three elements. We'll call this list *lu*.

In[5]:= **Clear[lu]**
lu = LUDecomposition[A]

Out[6]= {{{1, 2, 1}, {1, 1, 3}, {2, 3, -3}}, {1, 2, 3}, 1}

The first element is *Mathematica*'s special combination of upper and lower triangular matrices, Note that this is not the same as the L and U that you would compute by hand.

138

3.5 Mathematica On Line.nb

In[7]:= `lu[[1]]`

$$Out[7]= \begin{pmatrix} 1 & 2 & 1 \\ 1 & 1 & 3 \\ 2 & 3 & -3 \end{pmatrix}$$

LUDecompositon works with matrices that need row interchanges in order to find a LU decomposition. It also uses a process called partial pivoting in order to maintain numerical stability in the computations. The second element of the list is a vector specifying rows used for pivoting.

In[8]:= `lu[[2]]`

Out[8]= {1, 2, 3}

Notice that there have been no row interchanges here. Finally, if the matrix contains exact elements the third element of the list will be 1. With approximate numerical matrices, the third element of the list will be an approximation to the L^∞ condition number. In our case, the matrix is exact so we obtain 1 and the third element.

In[9]:= `lu[[3]]`

Out[9]= 1

We can change our matrix from exact to numerical with the function *N*. Compare the following numerical results with our previous exact results.

In[10]:= `luNumerical = LUDecomposition[N[A]]`

Out[10]= {{{2., 7., 8.}, {0.5, -1.5, -3.}, {0.5, 0.333333, 1.}},
{3, 1, 2}, 64.9306}

This time we see that there is a different intermediate matrix, there has been pivoting, and the condition number approximation is close to 65.

The second step in solving a linear system $A \cdot x = b$, is back substitution. *LUBackSubstitution* performs this operation.

In[11]:= **solution = LUBackSubstitution[lu, b]**

Out[11]= $\{\frac{7}{3}, -1, \frac{2}{3}\}$

We can check our result.

In[12]:= **A . solution**

Out[12]= {1, 2, 3}

We get equivalent results with our numerical representation.

In[13]:= **solutionNumerical = LUBackSubstitution[luNumerical, b]**

Out[13]= {2.33333, -1., 0.666667}

In[14]:= **A . solutionNumerical**

Out[14]= {1., 2., 3.}

The matrix returned by *LUDecompositionr* may be inscrutable; however, if the list is passed on to *LUBackSubstitution* the correct answer from backsubstitution will be given. On one hand this is annoying because we cannot easily see L and U. On the other hand, we don't have to worry about row interchanges and the backsubstitution is easily done with *LUBackSubstitution*. Mathematica has opted to work efficiently with the matrices so that we can find our answers easily rather than give us all of the intermediate steps. If you want the LU decomposition of a matrix where you can easily identify L and U, you will have to do the calculation by hand.

■ Exerices

1. Here is the matrix *A* from Example 1, Section 3.5 of your text.

In[15]:= **Clear[A]**
        A = {{1, 2, 1}, {2, 1, 1}, {1, 3, 2}}

Out[16]= $\begin{pmatrix} 1 & 2 & 1 \\ 2 & 1 & 1 \\ 1 & 3 & 2 \end{pmatrix}$

3.5 Mathematica On Line.nb

Here is a matrix b.

In[17]:= **b = {2, -1, 3}**

Out[17]= {2, -1, 3}

Use *LUDecomposition* and *LUBackSubstitution* to solve this system. Check your answer.

2. For additional vectors b_2 and b_3 of your choice use *LUBackSubstitution* to solve the corresponding systems of equations. (Note: you do not need *LUDecomposition* to perform this computation, just *LUBackSubstitution*.) Check your answers by computing $A \cdot x_2$ and $A \cdot x_3$.

3. We can compute L and U by hand. This is how you get started.

In[18]:= **Clear[U]**
U = A

$$Out[19]= \begin{pmatrix} 1 & 2 & 1 \\ 2 & 1 & 1 \\ 1 & 3 & 2 \end{pmatrix}$$

In[20]:= **U[[2]] = U[[2]] - 2 U[[1]];**
U[[3]] = U[[3]] - 1 U[[1]];
U

$$Out[22]= \begin{pmatrix} 1 & 2 & 1 \\ 0 & -3 & -1 \\ 0 & 1 & 1 \end{pmatrix}$$

Continue this process to obtain the upper triangular matrix U. From your computations, determine the lower triangular matrix L and check that the product $L \cdot U = A$.

On Line
4.1 Coordinates

We want our matrices to appear in standard form.

In[1]:= **$PrePrint = If[MatrixQ[#], MatrixForm[#], #] &**

Out[1]= If[MatrixQ[#1], #1, #1]&

■ Discussion

■ Parametric Equations for Ellipses and Hyperbolas

We can graph the ellipse whose equation is $\frac{x^2}{3^2} + \frac{y^2}{2^2} = 1$ by using the following parametric equations.

In[2]:= **Clear[x, y, t]**
  x[t_] = 3 Cos[t]
  y[t_] = 2 Sin[t]

Out[3]= 3 Cos[t]

Out[4]= 2 Sin[t]

Now we can use *ParametricPlot* to provide the graph of the ellipse. We need the *AspectRatio* so that the scale on the *x*- and *y*-axes is the same. Otherwise, the graph would be skewed in one direction or another.

4.1 Mathematica On Line.nb

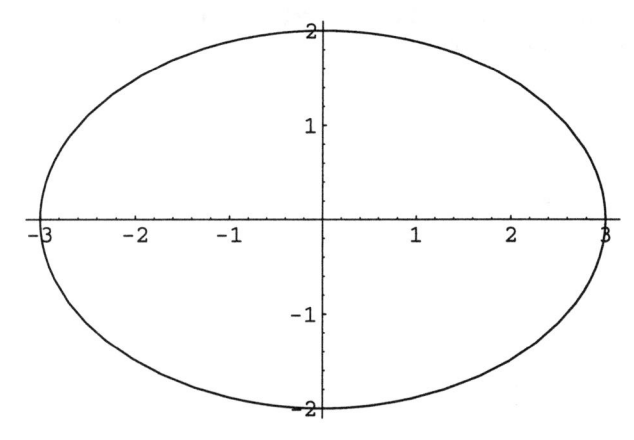

In[5]:=
```
ParametricPlot[{x[t], y[t]}, {t, 0, 2Pi},
    AspectRatio->Automatic]
```

Out[5]= - Graphics

These parametric equations give us this graph because they satisfy the ellipse's equation.

In[6]:= `Simplify[x[t]^2/3^2+y[t]^2/2^2]`

Out[6]= 1

We can also graph the hyperbola whose equation is given by $\frac{x^2}{3^2} - \frac{y^2}{2^2} = 1$ by using the following parametric equations. We can plot both the right and left sides of the hyperbola separately.

```
In[7]:= Clear[x, y, t]
        x[t_] = 3 Cosh[t]
        y[t_] = Sinh[t]
        rightSide = ParametricPlot[{x[t], y[t]}, {t, -2, 2},
            AspectRatio->Automatic]
```

Out[8]= 3 Cosh[t]

Out[9]= Sinh[t]

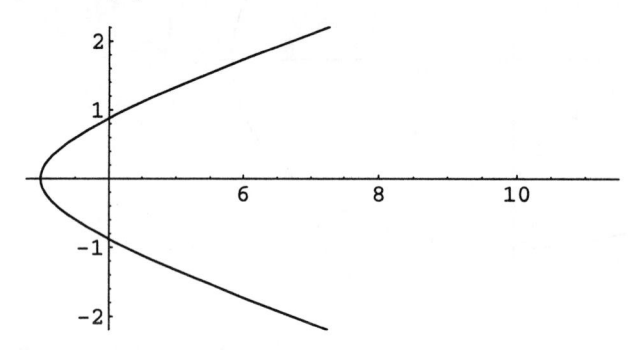

Out[10]= - Graphics

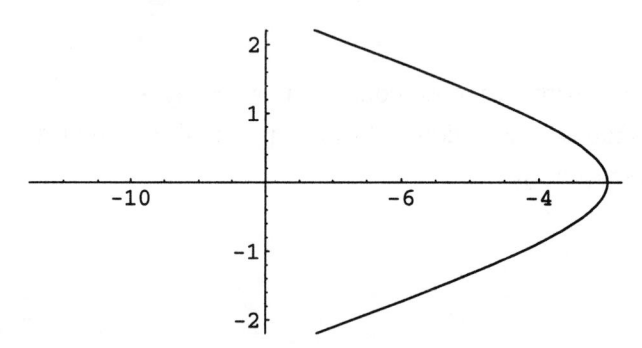

```
In[11]:= leftSide = ParametricPlot[{-x[t], y[t]}, {t, -2, 2},
            AspectRatio->Automatic]
```

Out[11]= - Graphics

We can show both branchs of the hyperbola at the same time.

In[13]:= **Show[leftSide, rightSide]**

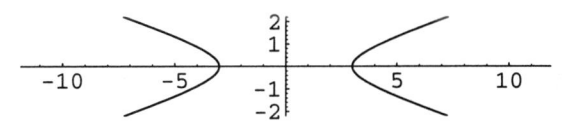

Out[13]= - Graphics

This hyperbola has asymptotes $y = \pm \frac{x}{3}$.

In[14]:= **asymptotes = Plot[{x/3, -x/3}, {x, -10,10}]**

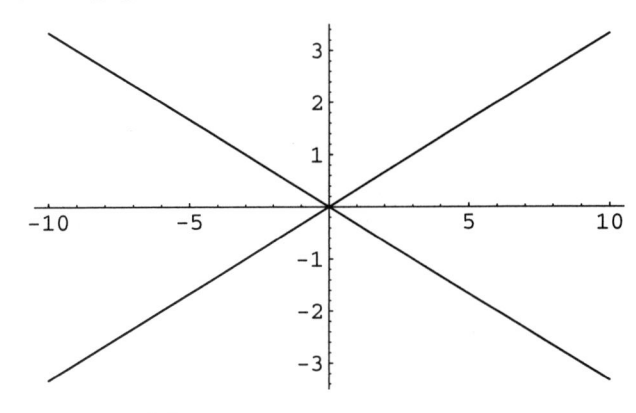

Out[14]= - Graphics

Finally, we can put this altogether in one graph.

In[15]:= **Show[leftSide, rightSide, asymptotes, AspectRatio->Automatic]**

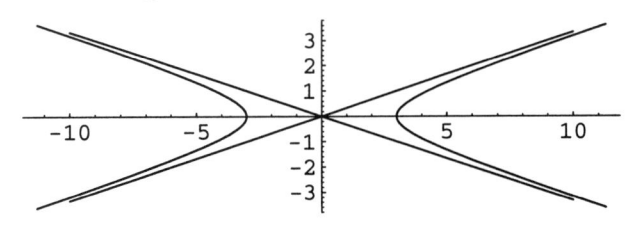

Out[15]= - Graphics

Change of Coordinates.

We want to use the asymptotes to form a new set of axes. We will find unit vectors pointing in the direction of each of the asymptotes. The points {3, -1} and {3, 1} both lie on the asymptotes. We find unit vectors as follows:

In[16]:= `Clear[q1, q2]`
`q1 = {3, -1} / Sqrt[3^2 + (-1)^2]`
`q2 = {3, 1} / Sqrt[3^2 + 1^2]`

Out[17]= $\{\frac{3}{\sqrt{10}}, -\frac{1}{\sqrt{10}}\}$

Out[18]= $\{\frac{3}{\sqrt{10}}, \frac{1}{\sqrt{10}}\}$

We can find the point matrix which has q_1 and q_2 as column vectors by taking the transpose of the matrix with those vectors as rows.

In[19]:= `pointMatrix = Transpose[{q1, q2}]`

Out[19]= $\begin{pmatrix} \frac{3}{\sqrt{10}} & \frac{3}{\sqrt{10}} \\ -\frac{1}{\sqrt{10}} & \frac{1}{\sqrt{10}} \end{pmatrix}$

The coordinate matrix is the inverse of the point matrix.

In[20]:= `coordinateMatrix = Inverse[pointMatrix]`

Out[20]= $\begin{pmatrix} \frac{\sqrt{\frac{5}{2}}}{3} & -\sqrt{\frac{5}{2}} \\ \frac{\sqrt{\frac{5}{2}}}{3} & \sqrt{\frac{5}{2}} \end{pmatrix}$

A point on the hyperbola can be found by evaluating *x* and *y* at any value of *t*.

In[21]:= `regCoords = {x[1.0], y[1.0]}`

Out[21]= {4.62924 1.1752}

We can find new coordinates for this point by multiplying by the coordinate matrix.

4.1 Mathematica On Line.nb

In[22]:= **skewCoords = coordinateMatrix . regCoords**

Out[22]= {0.58166, 84.2979}

We can define new parametric equations with respect to the skew coordinates.

In[23]:= **{xSkew[t_], ySkew[t_]} =
        coordinateMatrix . {x[t], y[t]}**

General::spell1 : Possible spelling error: new symbol
    name "ySkew" is similar to existing symbol "xSkew".

Out[23]= $\{\sqrt{\frac{5}{2}} \text{Cosh}[t] - \sqrt{\frac{5}{2}} \text{Sinh}[t], \sqrt{\frac{5}{2}} \text{Cosh}[t] + \sqrt{\frac{5}{2}} \text{Sinh}[t]\}$

We can graph the hyperbola with respect to the skewed coordinates by graphing the skewed parametric equations.

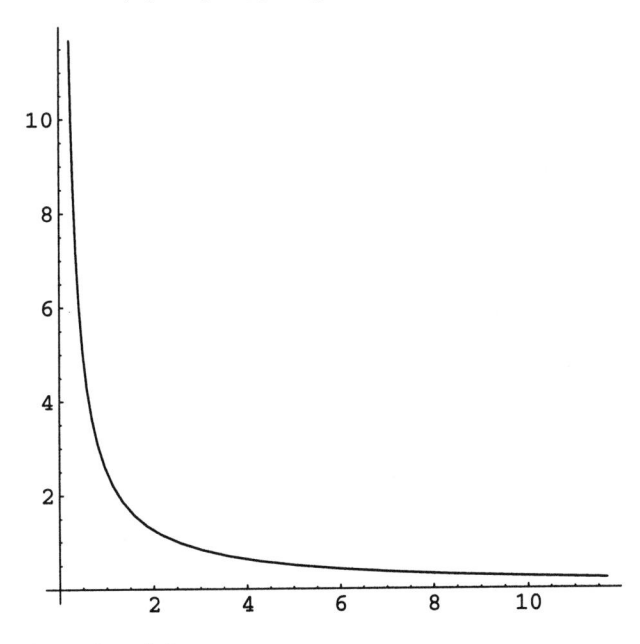

In[24]:= **ParametricPlot[{xSkew[t], ySkew[t]},
        {t, -2, 2}, AspectRatio->Automatic]**

Out[24]= - Graphics

This graph looks like graphs of equations of the form *xSkew y Skew* = constant. We can find the constant as follows:

In[25]:= **xSkew[t] ySkew[t]**

Out[25]= $\left(\sqrt{\frac{5}{2}} \text{Cosh}[t] - \sqrt{\frac{5}{2}} \text{Sinh}[t]\right) \left(\sqrt{\frac{5}{2}} \text{Cosh}[t] + \sqrt{\frac{5}{2}} \text{Sinh}[t]\right)$

We can multiply out this product.

In[26]:= **Expand[xSkew[t] ySkew[t]]**

Out[26]= $\frac{5 \text{Cosh}[t]^2}{2} - \frac{5 \text{Sinh}[t]^2}{2}$

Because of the identity $\text{Cosh}[t]^2 - \text{Sinh}[t]^2 = 1$, we see that the constant is 5/2.

■ Exercises

1. Consider the hyperbola whose equation is $x^2 - \frac{y^2}{2^2} = 1$.

(a) Graph the hyperbola, along with the appropriate aymptotes on a single set of axes.

(b) Let the asymptotes determine a new set of skewed axes. Find four points, two on each branch of the hyperbola, and find the skewed coordinates of these points. Multiply together the skewed coordinates of each of these points. Do you obtain the same result each time? Explain.

(c) Graph the hyperbola with respect to the new skewed axes.

(d) What is the new equation for the hyperbola in terms of the skewed coordinates?

2. Consider the ellipse $\frac{x^2}{4^2} - \frac{y^2}{2^2} = 1$.

(a) Graph the ellipse.

(b) Graph the ellipse after it has been rotated by 45 degrees.

(c) Graph the ellipse after it has been rotated by 30 degrees and translated 3 units in the x direction and 2 units in the y direction.

On Line
4.2 Projections

We want our matrices to appear in standard form.

In[1]:= **$PrePrint = If[MatrixQ[#], MatrixForm[#], #] &**

Out[1]= If[MatrixQ[#1], #1, #1]&

We will use the *GramSchmidt* function from the *Orthogonalization* package.

In[2]:= **<<LinearAlgebra`Orthogonalization`**

■ Discussion

You are working for an engineering firm and your boss insists that you find one single solution to the following system:

$$2x + 3y + 4z + 3w = 12.9$$
$$4x + 7y - 6z - 8w = -7.1$$
$$6x + 10y - 2z - 5w = 5.9$$

You object, noting that

(a) The system is clearly inconsistent: the sum of the first two equations contradicts the third.

(b) You need at least four equations to uniquely determine four unknowns. Even if the system were solvable, you couldn't produce just one solution.

    The boss won't take no for an answer. Concerning (a), the boss points out that the system was obtained from measured data and any inconsistencies can only be due to experimental error. Indeed, if any one of the constants on the

4.2 Mathematica On Line.nb

right sides of the equations were modified by 0.1 unit in the appropriate direction, the system would be consistent.

Concerning (b), the boss says, "Do the best you can. We will pass this data on to our customers and they wouldn't know what to do with multiple answers."

After some thought, you realize that projections can help with the inconsistency problem. The above system can be written in vector form as

$$x\begin{pmatrix}2\\4\\6\end{pmatrix} + y\begin{pmatrix}3\\7\\10\end{pmatrix} + z\begin{pmatrix}4\\-2\\-2\end{pmatrix} + w\begin{pmatrix}3\\-8\\-5\end{pmatrix} = \begin{pmatrix}12.9\\-7.1\\5.9\end{pmatrix}$$

(You can think of the first four vectors as column vectors of the matrix of coefficients and the fifth vector as the vector of right hand side constants.) You realize that this system would be solvable if the vector on the right of the above equality were in the space spanned by the four vectors on the left. Here is the matrix of coefficients.

$In[3]:=$ **Clear[A]**
A = {{2, 3, 4, 3}, {4, 7, -6, -8},
{6, 10, -2, -5}}

$Out[4]= \begin{pmatrix}2 & 3 & 4 & 3\\4 & 7 & -6 & -8\\6 & 10 & -2 & -5\end{pmatrix}$

We can row reduce A.

$In[5]:=$ **RowReduce[A]**

$Out[5]= \begin{pmatrix}1 & 0 & 23 & \frac{45}{2}\\0 & 1 & -14 & -14\\0 & 0 & 0 & 0\end{pmatrix}$

From this it is clear that the rank of A is 2 and, hence, the space spanned by the columns of A is a two-dimensional plane. (Call this plane W.)

Your idea is to let p be the projection of $b = \{12.9, -7.1, 5.9\}$ onto W. Since the system is so nearly consistent, p should be very close to b. Furthermore, the system $A \cdot x = p$ should certainly be solvable and one of the solutions should be what the boss is looking for.

Your point (b) will require some further thought. However, you do eventually come up with an idea which will be described in the exercises below.

■ Exercises

1. Here are the column vectors of A. (Notice that we are writing these as vectors rather than 3×1 column matrices.)

```
In[6]:= Clear[c1, c2, c3, c4]
        c1 = Transpose[A][[1]]
        c2 = Transpose[A][[2]]
        c3 = Transpose[A][[3]]
        c4 = Transpose[A][[4]]

Out[7]= {2, 4, 6}

Out[8]= {3, 7, 10}

Out[9]= {4, -6, -2}

Out[10]= {3, -8, -5}
```

Compute an orthonormal basis for the span of these column vectors. You could use the Gram-Schmidt process to compute the results or you can use the function *GramSchmidt* from the linear algebra package *Orthogonalization*. This function should be applied to a basis, and we see from the row-reduced form of A that the first two columns of A form a basis for the column space. Thus, we will apply *GramSchmidt* to c_1 and c_2. The output is a matrix whose rows are orthonormal vectors that span the space W.

4.2 Mathematica On Line.nb

```
In[11]:= {q1, q2} = GramSchmidt[{c1, c2}]
```

$$Out[11] = \begin{pmatrix} \frac{1}{\sqrt{14}} & \sqrt{\frac{2}{7}} & \frac{3}{\sqrt{14}} \\ -\frac{5}{\sqrt{42}} & 2\sqrt{\frac{2}{21}} & -\frac{1}{\sqrt{42}} \end{pmatrix}$$

Let's check the orthonormality.

```
In[12]:= q1 . q1
         q2 . q2
         q1 . q2

Out[12]= 1

Out[13]= 1

Out[14]= 0
```

Here is the vector b, the right hand side of our system of equations.

```
In[15]:= b = {12.9, -7.1, 5.9}

Out[15]= {12.9, -7.1, 5.9}
```

Use q_1 and q_2, and the Fourier Theorem from the text, to find the projection b_0 such that the vector $b_1 = b - b_0$ is perpendicular to W.

Finally, apply *LinearSolve* and *NullSpace* to find all solutions to $A \cdot x = b_0$. Express your solution in parametric form. The output of *LinearSolve* will be the translation vector and the output of *NullSpace* will give you vectors that should be multiplied by parameters.

2. Concerning your objection (b), your first thought is that maybe it would be reasonable just to report the translation vector as the soluton. But there is nothing special about the translation vector. The Translation Thoerem says that the general solution can be expressed using any particular solution instead of the translation vector.

Your next idea, however, is a good one. Let t be the translation vector and let t_0 be its projection to the nullspace of A. Let $x = t - t_0$. Then x is what you will report to your boss. Why is x a solution?

Find x.

3. Try computing x starting with some solution other than t. You should get the same x. Why does it work out this way?

Remark: It can be shown that your x is the solution of minimal length.

4. Although your solution was ingenious, *Mathematica* is "way ahead of you." Enter the following expression and compare the result with your answer x. This solution is called the "pseudo-inverse" solution. We could devote a whole chapter to it alone!

```
In[16]:= PseudoInverse[A] . b
Out[16]= {0.661409, 0.988255, 1.37684, 1.04613}
```

5. After giving the boss your answer, you delete all of your data except for A and x. A month later the customer calls, saying, "We know that there must be other solutions. Could you please provide us with the general solution?"

You are able to provide the desired information immediately, simply by entering one single *Mathematica* expression. How?

On Line
4.3 Fourier Series: Scalar Product Spaces.

■ Discussion

We can define the "rasp" function from the text using the following conditional definition. Note that the symbol ":=" must be used in place of the standard symbol "=". This results in a delayed definition and is appropriate here because mathematica must first know the value of x before deciding which conditional definition of *rasp* to use. This function will only be defined for the interval from -3 to 3.

```
In[1]:= Clear[rasp, x]
    rasp[x_] := x + 2 /; -3 <= x <= -1
    rasp[x_] := x     /; -1 <  x <= 1
    rasp[x_] := x - 2 /;  1 <  x <= 3
```

We can plot this function. We'll give it a label so that we can display it again below.

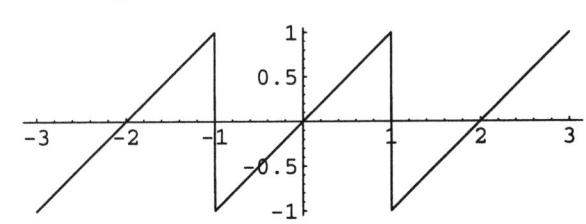

```
In[5]:= Clear[raspPlot]
    raspPlot = Plot[rasp[x], {x, -3, 3},
        AspectRatio->Automatic]
```

Out[6]= - Graphics -

We can compute the Fourier coefficients by integrating $x\, Sin[x]$ over the interval from -1 to 1.

155

$In[7]:=$ **Clear[a, n]**
  a[n_] = Integrate[x Sin[n Pi x], {x, -1,1}]

$Out[8]=$ $-\dfrac{2\cos[n\pi]}{n\pi} + \dfrac{2\sin[n\pi]}{n^2\pi^2}$

We can find the third approximation to the *rasp*.

$In[9]:=$ **Clear[f3]**
  f3[x_] = Sum[a[n] Sin[n Pi x], {n, 1, 3}]

$Out[10]=$ $\dfrac{2\sin[\pi x]}{\pi} - \dfrac{\sin[2\pi x]}{\pi} + \dfrac{2\sin[3\pi x]}{3\pi}$

Now we can graph the approximation.

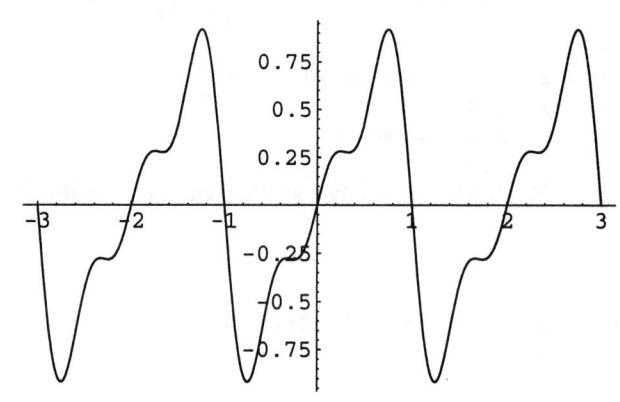

$In[11]:=$ **Clear[approxPlot3]**
  approxPlot3 = Plot[f3[x],{x,-3,3}]

$Out[12]=$ - Graphics -

4.3 Mathematica On Line.nb

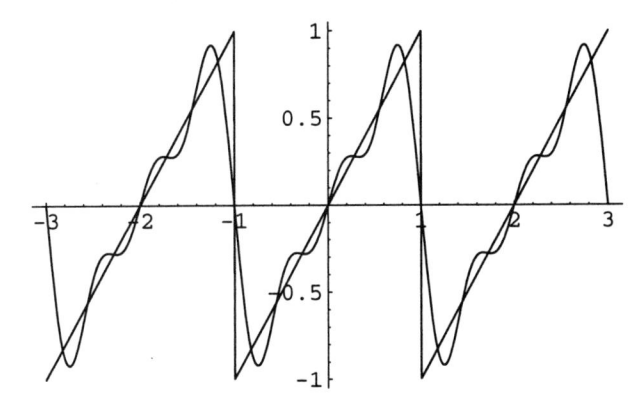

In[13]:= **Show[approxPlot3,raspPlot]**

Out[13]= - Graphics -

■ Exercises

1. Plot the first, fourth and tenth approximations to *rasp*.

2. Here is a definition of a *squareWave* function.

```
In[14]:= Clear[squareWave]
        squareWave[x_] := -1  /; -3 <= x <= -2
        squareWave[x_] :=  1  /; -2 <  x <  -1
        squareWave[x_] := -1  /; -1 <= x <=  0
        squareWave[x_] :=  1  /;  0 <  x <=  1
        squareWave[x_] := -1  /;  1 <  x <=  2
        squareWave[x_] :=  1  /;  2 <  x <=  3
```

Here is a plot of this function.

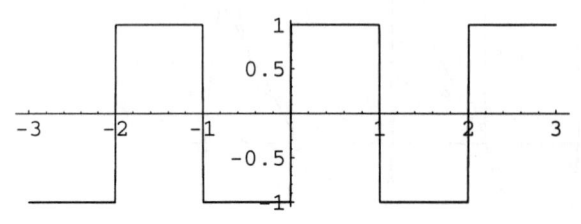

```
In[21]:= Clear[squareWavePlot]
         squareWavePlot = Plot[squareWave[x],
              {x,-3,3}, AspectRatio->Automatic]
```

Out[22]= - Graphics -

We can compute the Fourier coefficients *a*[*n*] as follows. We need to use two integrations because *Mathematica* won't integrate with conditional functions.

```
In[23]:= Clear[a, n]
         a[n_] = Integrate[-1 Sin[n Pi x],{x, -1, 0}] +
                 Integrate[ 1 Sin[n Pi x],{x, 0, 1}]
```

$$Out[24]= \frac{2}{n\pi} - \frac{2\cos[n\pi]}{n\pi}$$

Compute and graph the best approximations to the square wave function showing each approximation and the square wave function on the same set of axes.

Notice the "ear-like" peaks in the approximations at the discontinuities of the wave. These peaks are referred to as the "Gibbs phenomenon." They are quite pronounced, even after twenty terms of the Fourier series. Their existence shows that it takes a very high fidelity amplifier to accurately reproduce a square wave. For this reason, square waves are sometimes used to test the fidelity of an amplifier.

3. Consider the following function.

```
In[25]:= Clear[f, x]
         f[x_] := ( x + 2)^2    /; -3 <= x < -1
         f[x_] := x^2           /; -1 <= x <  1
         f[x_] := (x - 2)^2     /;  1 <= x <= 3
```

Here is its graph.

4.3 Mathematica On Line.nb

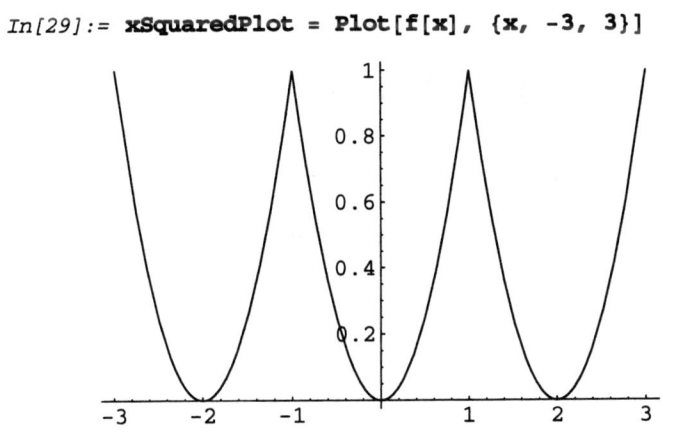

```
In[29]:= xSquaredPlot = Plot[f[x], {x, -3, 3}]
```

[Plot of a periodic function resembling x^2 on intervals, from -3 to 3, with peaks at integers and zeros at even integers]

```
Out[29]= - Graphics -
```

We can compute the Fourier coefficients as follows. Notice that we use x^2 rather than $f(x)$ because *Mathematica* won't integrate with functions defined conditionally.

```
In[30]:= Clear[a, n]
         a[n_] = Integrate[x^2 Sin[n Pi x], {x, -1, 1}]
```

```
Out[31]= 0
```

Explain why these coefficients make the sine approximations to f meaningless. Explain what the problem is.

4. Use cosine functions instead of sine functions for your approximations. Plot the approximations using one, four, and eight cosine functions and show the plots and the plot of f on the same set of axes.

On Line
4.4 Orthogonal Matrices

We want matrices to appear in standard form.

In[1]:= **$PrePrint = If[MatrixQ[#], MatrixForm[#], #] &**

Out[1]= If[MatrixQ[#1], #1, #1]&

■ Discussion

Let R_θ^x and R_θ^y be as defined in this section and define

$$A = R_{\pi/6}^x \cdot R_{\pi/4}^y$$

Since the product of two orthogonal matrices is orthogonal, A is orthogonal. The purpose of this exercise is to demonstrate that A defines a rotation around a fixed axis by a particular angle. Points on this axis remain fixed under the rotation. Thus, if x is on the axis of rotation, it will satisfy $A \cdot x = x$. Notice that this equation is equivalent with

$$(A - I) \cdot x = 0.$$

■ Exercises

1. Here are the matrices $R_{\pi/6}^x$ and $R_{\pi/4}^y$. In order to be more efficient in the computations that will follow, we will use numerical approximations for our entries rather than exact values.

4.4 Mathematica On Line.nb

```
In[2]:= Clear[RxPiOver6, RyPiOver4]
        RxPiOver6 = N[{{1, 0, 0}, {0, Cos[Pi/6], -Sin[Pi/6]},
           {0, Sin[Pi/6], Cos[Pi/6]}}]
        RyPiOver4 = N[{{Cos[Pi/4], 0, -Sin[Pi/4]}, {0, 1, 0},
           {Sin[Pi/4], 0, Cos[Pi/4]}}]
```

$$Out[3] = \begin{pmatrix} 1. & 0 & 0 \\ 0 & 0.866025 & -0.5 \\ 0 & 0.5 & 0.866025 \end{pmatrix}$$

$$Out[4] = \begin{pmatrix} 0.707107 & 0 & -0.707107 \\ 0 & 1. & 0 \\ 0.707107 & 0 & 0.707107 \end{pmatrix}$$

We can define A to be the product of these matrices.

```
In[5]:= Clear[A]
        A = RxPiOver6 . RyPiOver4
```

$$Out[6] = \begin{pmatrix} 0.707107 & 0. & -0.707107 \\ -0.353553 & 0.866025 & -0.353553 \\ 0.612372 & 0.5 & 0.612372 \end{pmatrix}$$

Check to see that A is orthogonal by computing the product of A and its transpose. Explain why the result indicates that A is orthogonal.

2. An $n \times n$ identity matrix can be given by the expression *IdentityMatrix*[n]. We can find an x on the axis of rotation with the following computation. (Here *NullSpace* returns a list containing a vector which is a basis for the nullspace. We use {x} instead of x so that x will be defined as a vector rather than a 3 by 1 matrix.) Evaluate the following expression.

```
In[7]:= Clear[x]
        {x} = NullSpace[A - IdentityMatrix[3]]
```

$$Out[8] = (\,-0.529904 \quad 0.819161 \quad 0.219493\,)$$

3. We want to plot the line segment from $-x$ to x. We need to create a graphics object which we will label *xx*. The line will be shown in red and its width will be 1/100 of the width of the display area. The function *Line* gives the line from $-x$ to x as a graphics primitive. We can display the graphics object *xx* by using the function *Show*.

In[9]:= **xx = Graphics3D[{RGBColor[1,0,0], Thickness[0.01],
        Line[{-x,x}]}]**

Out[9]= - Graphics3D -

In[10]:= **Show[xx]**

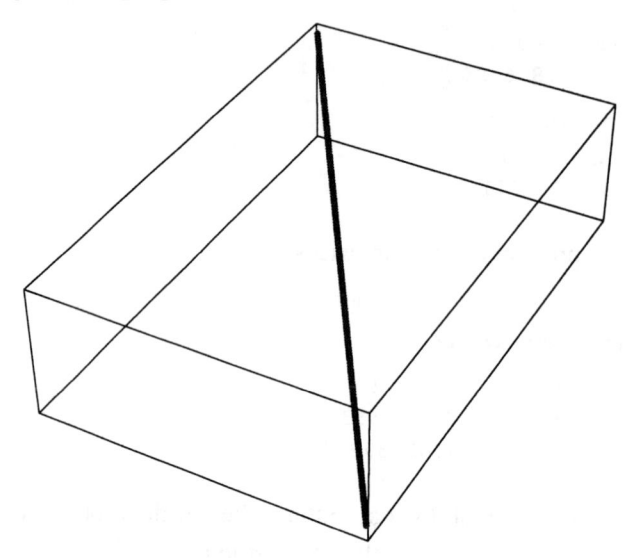

Out[10]= - Graphics3D -

4. The plane P through the origin perpendicular to x is called the "plane of rotation." Since this plane contains the origin, it is a subspace of R^3. It consists of all vectors y such that $x \cdot y = 0$. We can find a basis for P by finding the nullspace of the 3×1 matrix whose row is x. We define q_1 and q_2 to be the basis vectors for P

In[11]:= **Clear[q1, q2]
        {q1, q2} = NullSpace[{x}]//N**

Out[12]= $\begin{pmatrix} -0.848058 & -0.511848 & -0.137149 \\ 0. & -0.258819 & 0.965926 \end{pmatrix}$

Check to see that q_1 and q_2 are orthonomal by computing appropriate dot products.

4.4 Mathematica On Line.nb

We can "see" this basis by plotting the vectors q_1 and q_2 with their tails at the origin. We do this by plotting line segments from the origin to the tips of the vectors.

```
In[13]:= Clear[origin, basis]
         origin = {0, 0, 0}
         basis = Graphics3D[{RGBColor[0,1,0], Thickness[0.01],
            Line[{origin, q1}],Line[{origin, q2}]}]
```

Out[14]= {0, 0, 0}

Out[15]= - Graphics3D -

We can see both the axis, *xx*, and the basis vectors with the function *Show*.

```
In[16]:= Show[xx, basis]
```

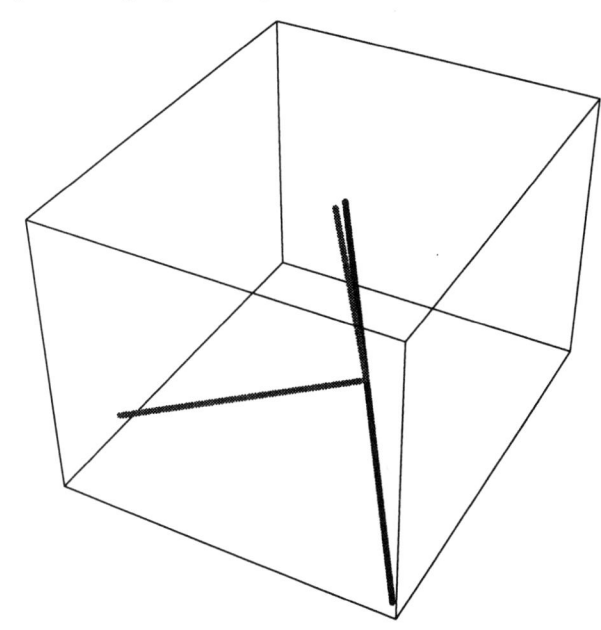

Out[16]= - Graphics3D -

From this angle, the basis elements appear neither orthogonal nor normal. To select a better viewing angle, we use the *ViewPoint* option to specify that we look at our vectors from the point {1, 3, 10}. Now it seems plausible that the basis is orthonormal.

```
In[17]:= Show[xx, basis, ViewPoint -> {1, 3, 10}]
```

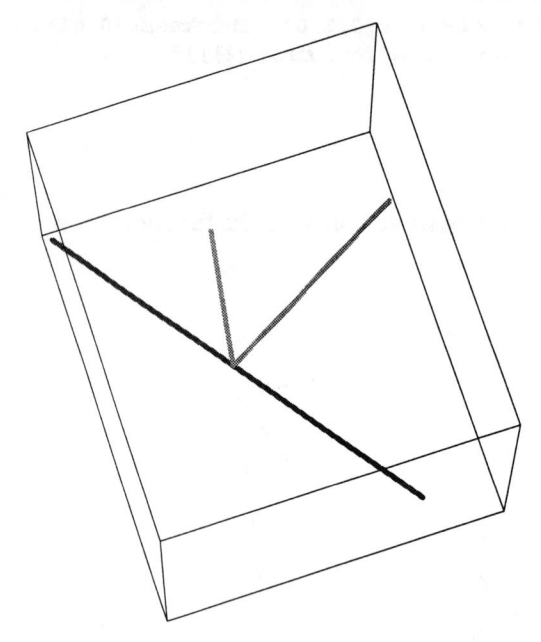

```
Out[17]= - Graphics3D -
```

5. We expect that multiplication by *A* should rotate elements in *P* around the axis of rotation by a fixed angle. To test this evaluate the following expressions. Explain what you see.

4.4 Mathematica On Line.nb   165

```
In[18]:= Clear[vectorAq1]
         vectorAq1 = Graphics3D[{RGBColor[0, 0, 1],
             Thickness[0.01], Line[{origin, A . q1}]}]
         Show[xx, basis, vectorAq1, ViewPoint -> {2, -2, 2}]

Out[19]= - Graphics3D -
```

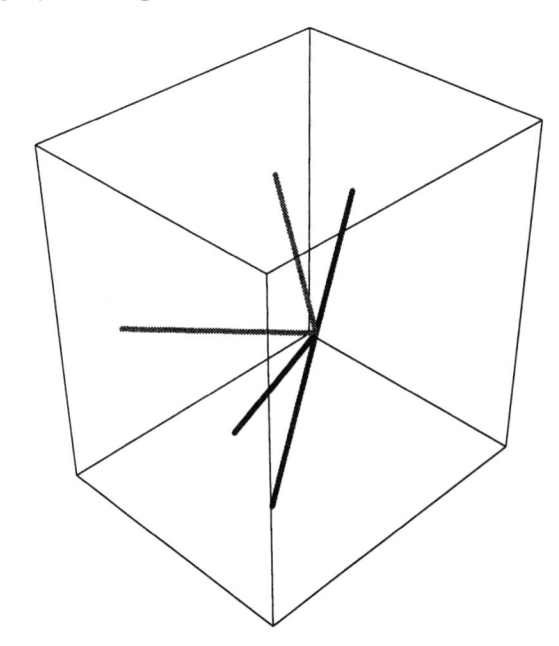

Out[20]= - Graphics3D -

6. We can look $A^n \cdot q_1$ for several values of n. To find the power of a matrix we must use the function *MatrixPower*. If we use the expression A^2, for example, we obtain a matrix each of whose entries is the square of the corresponding element of A. Here is an example.

```
In[21]:= Clear[B]
         B = {{1, 2}, {3, 4}};
         B^2
```

$$Out[23]= \begin{pmatrix} 1 & 4 \\ 9 & 16 \end{pmatrix}$$

We can plot $A^n \cdot q_1$ for values of n going from 1 to 15. Enter the following expressions and explain what you see.

```
In[24]:= manyVectors = Table[Graphics3D[{RGBColor[0,0,1],
            Thickness[0.01], Line[{origin,
            MatrixPower[A,n] . q1}]}],
            {n, 1, 15}]

Out[24]= {-Graphics3D-, -Graphics3D-, -Graphics3D-,
          -Graphics3D-, -Graphics3D-, -Graphics3D-,
          -Graphics3D-, -Graphics3D-, -Graphics3D-,
          -Graphics3D-, -Graphics3D-, -Graphics3D-,
          -Graphics3D-, -Graphics3D-, -Graphics3D-}

In[25]:= Show[xx, basis, manyVectors, ViewPoint->{1,3,10}]
```

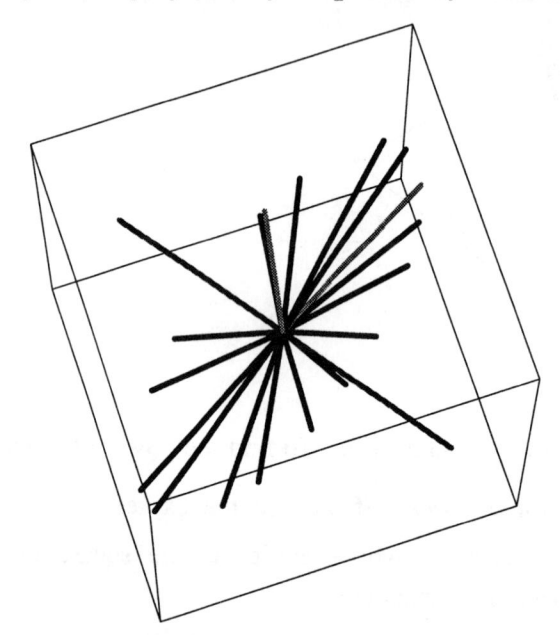

Out[25]= -Graphics3D-

6. Determine the angle of the rotation in P. (Hint: Use formula (4).)

On Line
4.5 Least Squares

We want our matrices to appear in standard form.

In[1]:= `$PrePrint = If[MatrixQ[#], MatrixForm[#], #] &`

Out[1]= `If[MatrixQ[#1], #1, #1]&`

■ Discussion

Given a polar equation, we can plot its graph using *ParametricPlot*. For example, in polar coordinates, an ellipse with one focus at the origin can be described by a formula of the form below where a, b, and c are constants.

$$r = \frac{c}{1 + a\sin\theta + b\cos\theta}$$

Here is the graph of one such ellipse. We use the *AspectRatio* option so that the scale on the x- and y-axes is the same.

```
In[2]:= Clear[r,t,x,y]
       r[t_] = 30 / (1  + 0.05 Sin[t] +0.8 Cos[t])
       x[t_] = r[t] Cos[t];
       y[t_] = r[t] Sin[t];
       ParametricPlot[{x[t], y[t]}, {t, 0, 2Pi},
           AspectRatio->Automatic]
```

$$Out[3]= \frac{30}{1 + 0.8\text{Cos}[t] + 0.05\text{Sin}[t]}$$

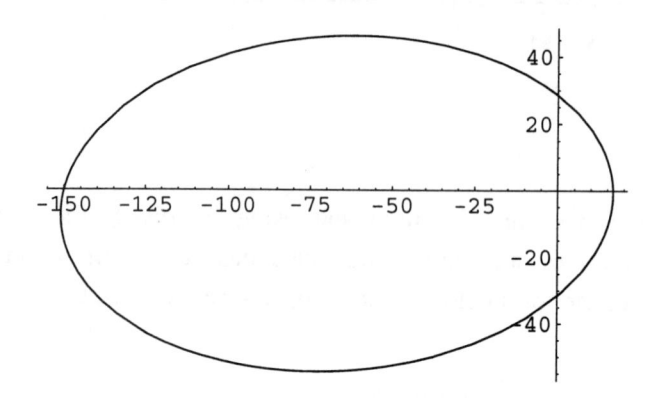

Out[6]= - Graphics

■ Exercises

1. Imagine that you are an astronomer who is investigating the orbit of a newly discovered asteroid. You want to determine (a) what is the closest the asteroid will come to the sun and (b) what is the furtherest away from the sun the asteroid will get. To solve your problem, you will make use of the following facts

(a) Asteroids have orbits which are approximately elliptical with the sun as one focus.

(b) You have collected the data below where r is the distance from the sun in millions of miles and θ is the angle between the vector from the sun to the asteroid and a fixed axis through the sun. This data is, of course, subject to experimental error. The following matrix contains the measured values of θ in the first column and the measured values of r in the second column.

4.5 Mathematica On Line.nb

```
In[7]:= Clear[thetaR]
        thetaR = {{0.00, 329.27}, {0.60, 313.80},
          {1.80, 319.49}, {1.40, 310.91}, {2.10, 327.88},
          {3.20, 372.91}, {5.40, 367.49}}
```

$$Out[8]= \begin{pmatrix} 0. & 329.27 \\ 0.6 & 313.8 \\ 1.8 & 319.49 \\ 1.4 & 310.91 \\ 2.1 & 327.88 \\ 3.2 & 372.91 \\ 5.4 & 367.49 \end{pmatrix}$$

Your strategy is to use the given data to find values of a, b, and c which cause the polar form of an ellipse, given above, to agree as closely as possible with the given data. This will involve setting up a system of linear equations and solving the normal equation. You will then graph the given formula and measure the desired data off of the graph. Good luck!

2. *Mathematica* has a way of solving least square problems which is easier (and better) than solving the normal equation. If the (inconsistent) system is expressed in the form $A . x = b$ then the least squares solution is *PseudoInverse*[A, b]. Try it!

On Line
5.1 Determinants

■ Discussion

It is only a slight exaggeration to say that the words "computer" and "determinant" are opposites. Determinants are extremely useful in many contexts. You will, for example, use them constantly when you study eigenvalues and eigenvectors later in the text. You will also see them used to write formulas for the solutions to many applied problems. In particular, determinants are used extensively in the study of differential equations and in the study of advanced calculus. Determinants are also used extensively in studying the mathematical foundations of linear algebra. However, if you ask a computer to find a numerical solution to any problem what-so-ever, it is a safe bet that the computer will not use determinants. Much faster and more efficient numerical techniques have been found. Thus, we will not provide any computer exercises for this chapter.

The reader should be aware, however, that *Mathematica* will computer determinants. Here is an example.

```
In[1]:= Clear[A, a, b, c, d]
        A = {{a, b}, {c, d}}
        Det[A]
```

$$Out[2] = \begin{pmatrix} a & b \\ c & d \end{pmatrix}$$

$Out[3] = -b c + a d$

On Line
6.1 Eigenvectors

We want our matrices to appear in standard form.

In[1]:= **$PrePrint = If[MatrixQ[#], MatrixForm[#], #] &**

Out[1]= If[MatrixQ[#1], #1, #1] &

■ Discussion

In the "On Line" section for Section 5.1, we commented that virtually anything that you might use determinants for, a computer would do otherwise. This includes finding eigenvalues. *Mathematica* uses a sophisticated computational technique which we shall not attempt to describe. In fact, it turns out that the algorithms for finding eigenvalues are so good that they often are used to find roots of polynomials! The exercises below explore this idea.

■ Exercises

1. We can show that the characteristic polynomial for the matrix $A = \begin{pmatrix} 0 & 1 \\ -b & -a \end{pmatrix}$ is $p(\lambda) = \lambda^2 + a\lambda + b$.

In[2]:= **Clear[A]**
       A = {{0, 1}, {-b, -a}}

Out[3]= $\begin{pmatrix} 0 & 1 \\ -b & -a \end{pmatrix}$

In[4]:= **Det[A - lambda IdentityMatrix[2]]**

Out[4]= b + a lambda + lambda^2

171

Use this to construct a matrix A which has $p(\lambda) = \lambda^2 + 7\lambda + 1$ as its characteristic polynomial. Use the *Mathematica* function *Eigenvalues[A]* to compute the eigenvalues of A and hence the roots of $p(\lambda)$. Check your calculation by using the quadratic formula to find the roots of $p(\lambda)$.

2. Compute the characteristic polynomial for the matrix

$$A = \begin{pmatrix} 0 & 1 & 0 \\ 0 & 0 & 1 \\ -c & -b & -a \end{pmatrix} \quad (4)$$

Use this information and the function *Eigenvalues* to approximate the roots of $p(\lambda) = \lambda^3 + 8\lambda^2 + 17\lambda + 10$. Test your answer by substituting your roots into $p(\lambda)$.

3. Here is the matrix which you used to approximate the roots of $p(\lambda)$ in the last exercise.

In[5]:= **Clear[A]**
    A = {{0, 1, 0}, {0, 0, 1}, {-10, -17, -8}}

Out[6]= $\begin{pmatrix} 0 & 1 & 0 \\ 0 & 0 & 1 \\ -10 & -17 & -8 \end{pmatrix}$

We can find eigenvalues and eigenvectors for A as follows:

In[7]:= **{values, vectors} = Eigensystem[A]**

Out[7]= {{-5, -2, -1}, {{1, -5, 25}, {1, -2, 4}, {1, -1, 1}}}

Eigensystem returns a list of two objects, the first is a list of the eigenvalues of A.

In[8]:= **values**

Out[8]= {-5, -2, -1}

The second is a matrix whose rows are the eigenvectors of A.

6.1 Mathematica On Line.nb

```
In[9]:= vectors
```

$$Out[9] = \begin{pmatrix} 1 & -5 & 25 \\ 1 & -2 & 4 \\ 1 & -1 & 1 \end{pmatrix}$$

The eigenvalues and eigenvectors are listed in corresponding order. For example, compare the following:

```
In[10]:= A . vectors[[1]]

Out[10]= {-5, 25, -125}

In[11]:= values[[1]] vectors[[1]]

Out[11]= {-5, 25, -125}
```

What do you notice about the entries of *vectors[[1]]*? Do the other eigenvectors of A exhibit a similar pattern? Use this observation to describe the eigenvectors for the matrix in formula (4) above in terms of the roots of the characteristic polynomial. Prove your answer.

4. Find a 4×4 matrix A whose characteristic polynomial is

$$p(\lambda) = \lambda^4 + 3\lambda^2 - 5\lambda + 7.$$

Then use the *Mathematica* function *Eigenvalues* to compute the roots. (If you use exact values in your matrix you may not be able to interpret the results. Try using decimals for the entries in your matrix in order to use numerical routines rather than exact routines in computing your answers.)

On Line
6.2 Diagonalization

We want our matrices to appear in standard form.

```
In[1]:= $PrePrint = If[MatrixQ[#], MatrixForm[#], #] &
Out[1]= If[MatrixQ[#1], #1, #1]&
```

■ Exercises

In this exercise, you will create your own eigenvalue problem. You should first enter a 3×3, rank 3, matrix P of your own choice into *Mathematica*. (You can use *RowReduce* to check the rank of your matrix.) To avoid trivialities, make each of the entries of P be non-zero. Next, enter a 3×3 *diagonal* matrix S whose diagonal entries are 2, 2, and 3 (in that order). Finally let $A = P \cdot S \cdot Inverse[P]$. The general theory predicts that

(a) Each column of P is an eigenvector for A and

(b) The eigenvalues of A are 2 and 3.

1. Verify (a) above by multiplying each column of P by A and checking to see that they are indeed eigenvectors corresponding to the stated eigenvalues.

2. Verify (b) above with the *Mathematica* function *Eigenvalues*.

3. *Mathematica* can also compute eigenvectors. Enter the expression *{values, vectors} = Eigensystem[A]*. This will produce a list: the first element is a list of the eigenvalues of A and the second element is a matrix whose rows are the corresponding eigenvectors of A. As in Exercise 1 above, check that the rows of *vectors* really are eigenvectors of A. The row of *vectors* which corresponds to the eigenvalue 3 should be a multiple of the third column of P. (Why?) Check

6.2 Mathematica On Line.nb

that this really is the case. How should the other two rows of vectors relate to the columns of *P*? Check that this really is true.

4. There is only one degree three polynomial with roots 2, 2, and 3 which has x^3 as its highest degree term. What is this polynomial? (Hint: Write it as a product of linear factors and then expand.) You can check your work with the *Mathematica* expression *Det[x IdentityMatrix[3] - A]*. How does this polynomial differ from the characteristic polynomial defined in your text?

On Line
6.3 Complex Eigenvalues

We want our matrices to appear in standard form.

In[1]:= `$PrePrint = If[MatrixQ[#], MatrixForm[#], #] &`

Out[1]= `If[MatrixQ[#1], #1, #1]&`

■ Exercises

1. Here is the matrix from Example 2.

In[2]:= `Clear[A]`
        `A = {{1, -3}, {1, 1}}`

Out[3]= $\begin{pmatrix} 1 & -3 \\ 1 & 1 \end{pmatrix}$

We can find the eigenvalues and eigenvectors of *A*.

In[4]:= `{values, vectors} = Eigensystem[A]`

Out[4]= $\{\{1 - I\sqrt{3}, 1 + I\sqrt{3}\}, \{\{-I\sqrt{3}, 1\}, \{I\sqrt{3}, 1\}\}\}$

This demonstrates that *Mathematica* "knows about" complex eigenvalues. Verify that these are eigenvalues and eigenvectors by computing *A* . *vectors*[[*i*]] and *values*[[*i*]] *vectors*[[*i*]], for *i* = 1, 2 and comparing the results.

2. Let *A* be an *n*×*n* matrix with complex entries. We say that *A* is Hermitian symmetric if $\overline{A^t} = A$. Give an example (reader's choice) of a 3×3 Hermitian matrix. Keep all entries non-zero and use as few real entries as possible. Enter your matrix into *Mathematica*. (Complex numbers such as 2 + 3*i* are entered into *Mathematica* as "2 + 3 I.") Then get *Mathematica* to find the eigenvalues. You should discover (remarkably) that they are all real!

You may want to use at least one decimal entry in your matrix in order to use numerical methods rather than exact methods. It will also be much quicker. Do you get very small imaginary parts on your eigenvalues? How do you explain that when the eigenvalues should be real?

3. Change one of the entries of the matrix A from the previous exercise so as to make it non-Hermitian. (This can be done simply by reassigning the value of, say, $A[[2,1]]$.) Now compute the eigenvalues. Are they still real?

4. In mathematics, the transpose of the conjugate of a matrix is usually denoted $A^* = \overline{A^t}$. In *Mathematica* we can obtain A^* by *Conjugate*[*Transpose*[A]]. Try this on your matrices from Problems 2 and 3.

5. Enter into *Mathematica* a 3×3, real, symmetric matrix A. Make as many of the entries of A as possible distinct. In *Mathematica*, let $B = IdentityMatrix[3] + I\,A$. Let $C = IdentityMatrix[3] - 2\,Inverse[B]$ and compute C . *Conjugate*[*Transpose*[C]] and *Conjugate*[*Transpose*[C]] . C. Can you prove that what you observe is always true? (Hint: Begin with the equality $B + B^* = 2\,I$ and multiply by B^{-1} on the left and $(B^*)^{-1}$ on the right.)

On Line
6.5 Quadratic Forms: Orthogonal Diagonalization

We want our matrices to appear in standard form.

In[1]:= **$PrePrint = If[MatrixQ[#], MatrixForm[#], #] &**

Out[1]= If[MatrixQ[#1], #1, #1]&

∎ Discussion

The purpose of this exercise set is to check graphically a few of the general principles of Section 6.5. We begin by graphing the quadratic variety

$$x^2 + xy + 2y^2 = 1.$$

To graph it, we use the *Mathematica* function *ContourPlot*. This function plots contours or level curves for a function $f(x, y)$. We give the plot a name so that we may use it again later. The *Contours* option specifies which level curves should be plotted. In this case we want to plot the contour corresponding to $f(x, y) = 1$. The *ContourShading* option is turned off so that we see just the contour line and not different levels of shading. In order to have a smoother curve we use the *PlotPoints* option to increase the number of points used to plot the curve. (The default is 15.)

6.5 Mathematica On Line.nb

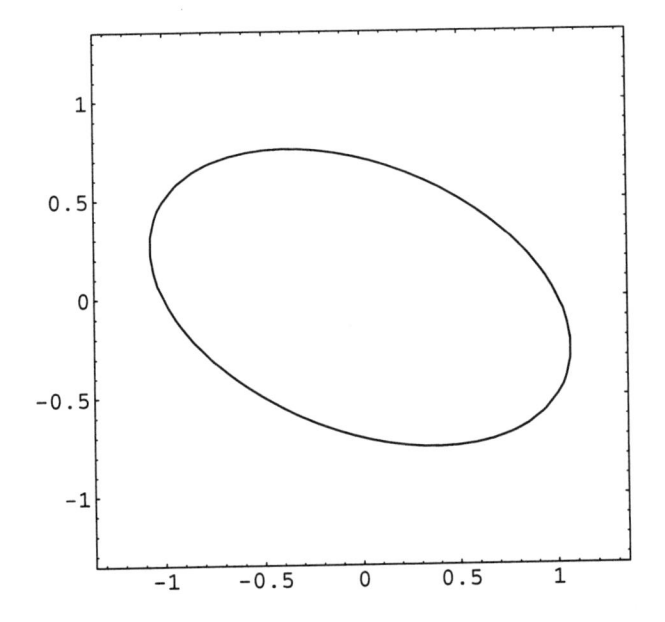

```
In[2]:= ellipse = ContourPlot[
          2 y^2 + x y + x^2, {x, -1.3, 1.3}, {y, -1.3, 1.3},
          ContourShading -> False,
          Contours -> {1},
          PlotPoints -> 30]
```

Out[2]= - ContourGraphics -

We define the matrix A corresponding to our quadratic variety.

```
In[3]:= Clear[A]
        A = {{1, 1/2}, {1/2, 2}}
```

$Out[4]= \begin{pmatrix} 1 & \frac{1}{2} \\ \frac{1}{2} & 2 \end{pmatrix}$

We can check that this gives the correct variety.

```
In[5]:= Expand[{x, y} . A . {x, y}]
```

$Out[5]= x^2 + x y + 2 y^2$

At this point we need to find the eigenvalues and eigenvectors for A.

$In[6]:=$ **Clear[values, vectors]**
   {values, vectors} = Eigensystem[A]

$Out[7]= \left\{\left\{\frac{1}{2}\left(3-\sqrt{2}\right), \frac{1}{2}\left(3+\sqrt{2}\right)\right\}, \{\{-1-\sqrt{2}, 1\}, \{-1+\sqrt{2}, 1\}\}\right\}$

We will want to normalize the eigenvectors so that they have length one. We define a function *norm* that finds the length of vectors.

$In[8]:=$ **Clear[norm, v]**
   norm[v_] = Sqrt[v . v]

$Out[9]= \sqrt{v . v}$

Now we can normalize our vectors.

$In[10]:=$ **vectors = {vectors[[1]] / norm[vectors[[1]]],**
   vectors[[2]] / norm[vectors[[2]]]}

$Out[10]= \begin{pmatrix} \frac{-1-\sqrt{2}}{\sqrt{1+(-1-\sqrt{2})^2}} & \frac{1}{\sqrt{1+(-1-\sqrt{2})^2}} \\ \frac{-1+\sqrt{2}}{\sqrt{1+(-1+\sqrt{2})^2}} & \frac{1}{\sqrt{1+(-1+\sqrt{2})^2}} \end{pmatrix}$

According to the general theory, the major and minor axes of the ellipse should lie along the coordinate axes determined by the eigenvectors for the matrix *A* which describes the ellipse. We'll define new axes in terms of the normalized eigenvectors. We create them as graphics objects so that we can show them along with the plot of the ellpise above.

$In[11]:=$ **Clear[newAxes]**
   newAxes = {Graphics[{RGBColor[1, 0, 0], Thickness[0.01],
      Line[{{0, 0}, vectors[[1]]}]}],
      Graphics[{RGBColor[0, 0, 1], Thickness[0.01],
      Line[{{0, 0}, vectors[[2]]}]}]}

$Out[12]= \{$ - Graphics -, - Graphics - $\}$

Now we can look at the ellipse and the new axes at the same time.

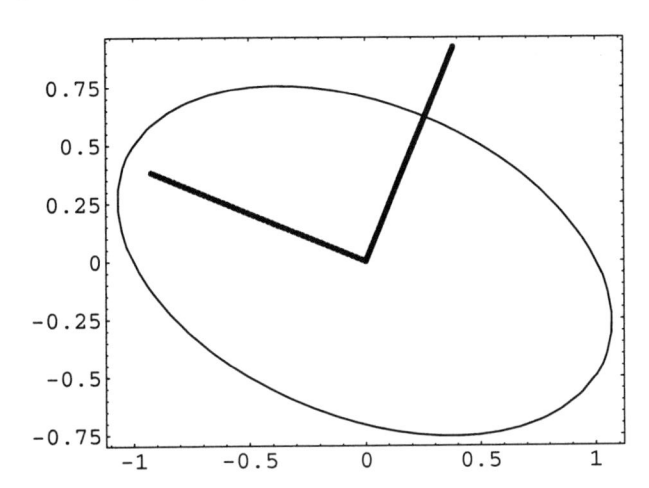

In[13]:= **Show[ellipse, newAxes, AspectRatio -> Automatic]**

Out[13]= - Graphics -

■ Exercises

1. According the the general theory, the ellipse should cross the new axes at the points $(\pm \frac{1}{\sqrt{\lambda_1}}, 0)$ and $(0, \pm \frac{1}{\sqrt{\lambda_2}})$. Verify this by modifying the definition of *newAxes* above by multiplying *vectors*[[i]] by *1/Sqrt*[*values*[[i]]] for $i = 1, 2$. After this modification you should be able to show both the ellipse and the axes and see that the lines represent the semi-major and semi-minor axes of the ellipse.

2. Find a formula for the ellipse centered at the origin for which

(a) the major axis is 4 units long and lies along the line determined by the vector $\{3, 4\}$ and

(b) the minor axis is 2 units long.

(Hint: If you can figure out what Q and S (diagonal matrix) should be, then you can set $A = Q \cdot S \cdot Q^t$.)

Get *Mathematica* to graph the ellipse as well as the semi-major and semi-minor axes. The axes should be determined by the eigenvectors for the corresponding symmetric matrix.

Maple On Line

Chapter 1

Systems

1.1 On Line

In this introductory section we will pose no exercises, but instead, will detail how to use Maple to solve problems in linear algebra. For the novice Maple user, this section is essential reading and reference. For the experienced Maple user, this section can be examined for evidence of any new tips and ideas for working with Maple. In addition, this section will reveal the working style guiding the solutions in the rest of this manual.

Maple is a symbolic, as well as a numeric, language. For users with experience in numerical computation (Matlab, BASIC, FORTRAN, C, etc.) there are differences in thinking that accompany calculating in Maple. The presence of symbolic variables adds a dimension missing in strictly numeric languages. For one thing, Maple is capable of computing answers in exact (symbolic) form, so round off error need not be an issue in didactic experiments. For another, it really helps to have a vision as to how to mix symbolic and numeric calculations, lest operations performed numerically impinge on the operations which will be later performed symbolically.

We illustrate some sample Maple calculations, beginning with simple arithmetic. As we present the Maple syntax for various calculations, we will explain the Maple Release 4 interface as it appears under Windows on a PC. Release 4 has been engineered to have the same functionality on all major platforms (Macintosh, PC, UNIX) so only local differences in file structures and networks might possibly differ.

Enter Maple syntax at the prompt, the ">" at the left edge of the screen.

```
>   2/3 + 5/7;
```

Lines entered into Maple must have terminal punctuation, typically, the semi-colon (;). Simple arithmetic is done automatically. A new prompt is generated upon execution of the command, and that follows upon pressing the ENTER key. Spaces are generally ignored by Maple, and are used here to improve readability.

```
>   x + 2*x;
```

Maple applies the rules of algebra to symbolic expressions, and some of the simplifications are immediate. Others must be requested by a variety of special commands.

Assignments are made using the two characters "colon" and "equal" (:=), two characters that must not be separated by a space!

```
>   f := x^2;
```

We have just entered an *expression*, or formula, and assigned it to the name "f." To evaluate this expression at x = 3, use the **subs** command to substitute x = 3 into f.

```
>   subs(x=3,f);
```

We next articulate Lopez's Large Law (see the article Tips for Maple Instructors, *MapleTech*, Vol. 3, NO. 2, 1996, published by *Birkhauser*) which states "never use on the left of an assignment a variable in use somewhere on the right." Thus, it is not a good idea to make the assignment x := x + 1. Although this is an extremely common construction in numeric languages, it is most unwise in a symbolic language.

Moreover, Lopez's Large Law precludes assigning values to the "working variables" x, y, z, etc. A strict distinction between the letters being used for "names" on the left, and "variables" on the right makes Maple more user-friendly.

Since we have demonstrated how to make an assignment to the name "f" we next show how to "erase" that assignment.

```
>   f;
>   f := 'f';
>   f;
```

First, "erasing" consists of assigning f to its "letter value" which is what the single quotes accomplish. Second, entering "f;" interrogates Maple for what it knows about that name. Maple echoes the contents of that name.

It is possible to create unfathomable Maple Worksheets. Violations of Lopez's Large Law can lead to such difficulties, and so can misunderstand-

1.1 Maple On Line

ing Maple's file management. For example, suppose Lopez's Large Law has been violated by

>   x := 3;

Much later in the Worksheet, the "variable" x is used symbolically, as in

>   f := x*sin(Pi*x);

What happened? Why didn't the formula $x \sin \pi x$ appear? Since x has been assigned the value 3, Maple computed $\sin(3\pi)$ and got 0.

>   x := 'x';

>   f;

Violating Lopez's Large Law indeed has consequences. It is not enough to "erase" x. Once the assignment to f is the number 0, it remains the number 0. Having a clear strategy for working with Maple prevents needless frustration.

The next "time-bomb" is more subtle. Suppose that f has the value 0 from the above calculations. And suppose that all record of the existence of f in the Worksheet is removed by deleting the appearance of f from the Worksheet. The letter "f" is still assigned the value 0. In fact, it's worse than that. If another blank Worksheet is opened via the menu options File/New, this new Worksheet will still have f assigned the value 0. This is because both Worksheets share the same memory, and what is known to Maple in one Worksheet is known to the other. On some platforms, with the right initialization of Maple, each Worksheet can have its own attached memory. Unless you know for sure how your copy of Maple is installed, it is best to assume that all Worksheets share a single memory, and always, under all circumstances, remember that merely removing the appearance of an assignment from a Worksheet does not remove that assignment from Maple's memory.

Suppose at this point in your experiments with a Worksheet you have excised the last input/output pair, and the bottom of your Worksheet no longer contains "the next prompt." How do you generate a new prompt? Place the cursor in an input line. Simultaneously press the keys CTRL and k to insert a prompt *above* the cursor, and CTRL and j to insert a new prompt *below* the cursor. If you examine the Insert menu, the "Execution Group" corresponds to "a new prompt."

Having toyed with the menu bar, observe the Help menu. The best advice a Maple user can receive is to begin with Help/Contents. The resulting document that opens is hyperlinked to all sorts of information about Maple, and it is left to the user to navigate through instructions on the interface, and on Maple itself.

However, to get help on a command whose name you know, you can use the question mark.

```
>   ?subs
```

Help commands don't need terminal punctuation. At the bottom of most help screens is a section of Examples. Look there first. Examples can be copied and pasted back into your Worksheet for experimentation.

The exercises for each section in this manual begin with the instruction to "restart Maple." This means to issue the **restart** command which clears all variables. This command does not erase anything visible in the Worksheet, so if a Worksheet has become confused and entangled, issuing a restart and then working from the top down re-executing all the entered commands, is a way of "beginning at the beginning" and retracing the thought process in the Worksheet.

```
>   restart;

>   f;
```

We next illustrate some algebraic simplifications.

```
>   q := 1/x + 1/y;

>   q1 := simplify(q);
```

Commands in Maple typically take parentheses around the argument or arguments. The letter "q" makes a handy label because it is easy to find on the keyboard. Re-assigning a new version of a name to itself is actually a violation Lopez's Large Law, and should be avoided. Thus,

```
>   q := simplify(q);
```

is not a syntax error, but is just plain bad worksmanship. If parts of the Worksheet are re-executed, which version of q is being referenced? Is it the unsimplified version or the simplified version? So, use unique names for each meaningful Maple output to avoid confusion when experimenting, since that usually requires editing, changing, re-executing, moving up and down throughout the Worksheet. If the same letter has multiple meanings throughout the Worksheet, chaos results.

Maple crashes. It is a fact of life that Maple, inexplicably and explicably, crashes. It is exceedingly frustrating to have spent an hour or more on an assignment in Maple, only to lose it all because Maple crashed. There is only one piece of advice that makes sense here, and that is "Save early, and save often." The first time a Worksheet is saved (File/Save or the "diskette"

1.1 Maple On Line

icon on the toolbar) Maple prompts for a file name and a destination for the saved file. Thereafter, clicking the "diskette" icon on the toolbar, or entering CONTROL S from the keyboard, saves work to that same file. Save early and save often. That advice cannot be repeated too frequently.

Maple is both a symbolic and a numeric language. Thus, there is a difference between 1 and 1.0 in Maple. The first is the exact integer 1, while the second is the decimal version of the number 1. Converting the exact symbolic representation of a number is done with the **evalf** (evaluate floating point) command.

>   q := 1/sqrt(2);

>   evalf(q);

Note that Maple immediately rationalizes $\frac{1}{\sqrt{2}}$. And note further that Maple can provide many more than the default 10 digits.

>   evalf(q,20);

The next thing useful to know about Maple is how to reference parts of answers it generates. Consider the following solution to a quadratic equation.

>   q := x^2 + 3*x + 1 = 0;

>   solve(q,x);

Maple has returned a *sequence* of two roots, which can be referenced if a tag had been assigned to the **solve** command. Thus, the better working strategy is

>   q1 := solve(q,x);

Now, the roots can be referenced by the bracket notation

>   q1[1];

>   q1[2];

Thus, there are three data structures Maple uses that are worth understanding. Maple uses *sequences*, *lists*, and *sets*. In a sequence items are separated by commas. A list is a sequence enclosed by square brackets: [a,b,c]. A set is a sequence enclosed by curly braces: {a,b,c}. The list preserves order and replicas. The set does not.

>   [a,b,a,a,c];

>   {a,b,a,a,c};

Each structure reflects valid mathematical usage, and Maple has commands for manipulating each data structure properly.

Although repetitive tasks can be implemented in Maple by copying and pasting input lines, a for-loop is the appropriate way to repeat similar instructions. In this manual, the for-loop is entered as a single input as follows.

```
>   for k from 1 to 3 do x.k := k^2; od;
```

All three input lines are connected to the one prompt by entering all lines but the last with SHIFT ENTER, rather than simply ENTER. In Release 4 this is now more aesthetic than practical, but if the three lines of code above are entered into Maple V Release 4 with just the ENTER key, the first line will generate a complaint about "incomplete", a complaint that disappears when the terminating **od** ("do" spelled backwards) is entered. In previous versions of Maple, failure to keep the lines of a for-loop together could lead to terrible consequences if changes were made to individual lines of the loop. In Release 4 this is no longer such a problem.

One advantage of the notation x1, x2, x3 is that such objects can be referenced collectively by

```
>   x.(1..3);
```

For large collections of similar objects this turns out to be a handy device for saving repetitive and tedious typing.

We will be concerned primarily with Maple's functionality in linear algebra. Maple's code is modularized, bundled into related groups called packages. There are some 32 packages in Release 4, all of which are present in every properly installed version of Maple. The command

```
>   ?packages
```

brings up the list of packages, and the names of the packages are hyperlinked to more information about the individual packages.

The package we will use most is the *linalg* package, itself containing more that 100 commands for manipulating vectors and matrices. The *linalg* package is "loaded" into Maple via the command

```
>   with(linalg):
```

Notice that the terminal punctuation here is the colon (:) which suppresses output. This package will be loaded for every exercise set, and we will want to suppress the listing of the more than 100 commands made present by this package.

First, we enter the matrix $A = \begin{bmatrix} 1 & 2 \\ 3 & 4 \end{bmatrix}$.

1.1 Maple On Line

```
>   A := matrix(2,2,[1,2,3,4]);
```

It appears easier to provide the **matrix** command with a single list of entries, letting Maple wrap them according to the dimensions given first. The alternative is to give the **matrix** command a list of lists, the sub-lists being the rows of the matrix. Thus,

```
>   matrix([[1,2],[3,4]]);
```

We will use Maple's **randmatrix** command to generate matrices at random. Maple's random number generator will generate the same sequence of random numbers each time it is initialized by starting (or restarting) Maple. The advantage here is that results are then reproducible, even if random matrices have been used. There is a way of setting the starting point for the random number generator, but that is not used in any of the exercises. The help file for **rand**, the random number generator, accessed by ?rand, will mention the global variable _seed which can be assigned values (a student's SSN?), thereby making unique assignments for each student. That is not done in these exercises.

```
>   B := randmatrix(2,2);
```

Matrix and vector arithmetic is most easily done by applying **evalm** (evaluate matrix) to the desired arithmetical commands.

```
>   2*A + 3*B - A^2;
>   evalm(2*A + 3*B - A^2);
```

In the first instance, merely the names are manipulated by Maple. In the second, the actual entries of the matrices are manipulated. There are times and places for each approach, but only the second is used in these exercises.

Next, we address matrix multiplication, a process that is known in mathematics to be noncommutative. Thus, for numbers, 2*3 = 6, but for matrices, A B rarely equals B A. Hence, Maple distinguishes between the use of "*" for commutative multiplication, and "&*" for the noncommutative multiplication of matrices. In Release 4 Maple will warn specifically that A*B for matrices must be changed to A &* B. In earlier versions, the user had to be prescient.

```
>   evalm(A * B);
>   evalm(A &* B); evalm(B &* A);
```

Surprisingly, vectors will take more discussion than matrices. First of all, enter the vector $\mathbf{V} = \begin{bmatrix} -2 \\ 5 \end{bmatrix}$ with the following syntax. In a newly installed copy of Maple V Release 4 the output will be as you see below.

> V := vector([-2,5]);

Throughout this manual the output will instead appear as

> V := vector([-2,5]);

Why the difference and how do we get Maple to render vectors as "column-like" rather than "row-like"? And are such vectors "column vectors" or "row vectors"?

First, no matter how we get Maple to print the vector, it is always a column vector.

Second, to get your Maple session to print the vectors as columns, enter the following instructions.

> with(share): readshare(pvac,system):

These two commands cause Maple to load, from its Share Library, a file called *pvac* (print-vector-as-column), the effect of which is to change the way vectors are printed. If, for some reason, this functionality is to be switched off, enter the command

> pvac := false:

> print(V);

> pvac:=true:

> print(V);

In addition to noting how to turn this display feature on and off, observe that it takes **print** (or **evalm**) to get Maple to display the contents of a vector or matrix.

For the adventurous, from outside Maple, examine the file structure of the Maple V4 directory. There is a sub-folder called *Share* in which the contents of the Share Library are stored. A further sub-folder, *System*, contains another sub-folder called *Pvac*. In the *Pvac* folder there is a file Pvac.mpl, a file containing the code for the display feature being discussed. If this file can be rendered as a pure text file, it can be made into an initialization file so that the code will load automatically every time Maple is launched. The author of this manual has had this code running as an initialization file in both Release 3 and Release 4, a span of more two years. It has worked perfectly and has never given any trouble whatsoever.

To obtain a text version of the file pvac.mpl launch a text editor such as Word, etc. From within the text editor, locate the file pvac.mpl and open it. Perform a Save As, save the file as "text", give it the name Maple.ini (for the PC; on the Macintosh, use MapleInit, and for UNIX, use .mapleinit).

1.1 Maple On Line

Be sure that your text editor does not add a hidden .txt or other such extension. Quit the editor, and drop the initialization file into the Lib sub-folder of the Maple V4 folder for the PC, drop it into the Maple V4 folder for the Mac, and experiment with where to put it on a UNIX system. Relaunch Maple and you will automatically have launched the *pvac* code.

To verify that a Maple vector behaves as a column vector, no matter what it looks like, try the following experiments.

```
>   print(A,V);
>   evalm(A &* V);
```

This is exactly the product you should obtain from the product A **V** done by hand, treating **V** as a column vector. Now ask Maple to convert **V** to a matrix.

```
>   VC := convert(V,matrix);
>   type(V,vector); type(VC,vector);
>   type(V,matrix); type(VC,matrix);
>   VC[1,1];
>   VC[1,2];
>   VC[2,1];
```

The matrix VC does not have a second column. It has two rows. If the vector **V** is converted to a matrix data structure, it gets converted to a 2 x 1 (column) matrix. Maple thinks of the vector **V** as a column thing, even if its default print style is to make it look like a row thing.

Moreover, the transpose of **V**, if converted to a matrix, has all the characteristics of a 1 x 2 (row) matrix.

```
>   VR := convert(transpose(V),matrix);
>   VR[1,1];
>   VR[1,2];
>   VR[2,1];
```

The matrix VR does not have a second row. Maple thinks of the transpose of V as a row thing, no matter how it prints it. In fact, it is a great tragedy that the Maple programmers have decided that the default output to the **transpose** command is

```
>   transpose(V);
```

The evidence has already been presented that Maple understands the transpose. It is sad, indeed, that so fine a program as Maple should have such anomalous behavior when displaying the transpose of a vector **V** that so obviously has the inherent properties of a column object.

Incidentally, this means that there is no way, per se, to enter a row vector into Maple. You enter a column vector, the default vector object, then transpose the column vector. And, yes, you live with not being able to see the display of the transpose as a row-like object.

Caution: It is tempting to sidestep this issue of row and column vectors with the belief that instead, row and column matrices will be used. This is not a good idea. There are commands in Maple that specifically demand vectors, not matrices. For example, if you had defined, not V, but VC, a column matrix, and wanted to live your Maple life with only matrices, you'd run afoul of

> dotprod(VC,VC);

So, you cannot live without vectors, and if you cannot live without vectors, you must then face the issue of row and column vectors. Sorry, but to reap the benefits of Maple you have to put up with a few quirks. Kind of like life in general.

There is one final issue to face about linear algebra in Maple. To perform operations on vectors one must map the operator onto the vector. For example, to simplify a vector, use the following syntax.

> V := vector([1/x+1/y, 1/x-1/y]);

> simplify(V);

Obviously, not the right syntax.

> map(simplify,V);

The operator **simplify** has to be mapped onto the vector.

As another example, if **V** is a function of t and you want its derivative, you use the following syntax.

> V := vector([sin(t),cos(t)]);

> diff(V,t);

Obviously, the wrong syntax.

> map(diff,V,t);

Additional parameters to the mapped operator go at the end.

Unfortunately, there are two exceptions to the rule "Map things onto vectors and matrices."

1.2 Maple On Line

```
>   V := vector([Pi,2*Pi]);
>   evalf(V);
>   map(evalf,V);
```

Well, that seems to work. Why raise that as an exception? The command evalf can take an integer as a second argument, changing the number of digits returned.

```
>   evalf(Pi,20);
```

But if you try that for the vector V, it fails.

```
>   map(evalf,V,20);
```

Nonsense is returned. The method that works is

```
>   evalf(op(V),20);
```

A second exception is substitution.

```
>   V := vector([x,x^2]);
>   subs(x=1,V);
>   map(subs,V,x=1);
```

The syntax that works is

```
>   subs(x=1,op(V));
```

Hence, the rule is "Map all operators except **subs** and **evalf**. For those, don't **map**, but use **op** around the vector. If you use **map**, don't use **op**. When **op** is needed, you don't use **map**."

1.2 On Line

The exercises of this section explore the concept of "span" by visualizing randomly generated members of the span of a set of vectors. Begin by loading both the *linalg* and *plots* packages.

```
>   with(linalg):  with(plots):
```

Exercise 1

Enter into Maple the following four points. For most purposes, points can be represented as lists, a data structure denoted by square brackets.

>   P1:=[1,1]; P2:=[1,-1]; P3:=[-1,1]; P4:=[-1,-1];

The *plots* package makes available a **pointplot** command that will plot a list of points. There are a number of options to this command that will vary the look of the graph, and by using the toolbars associated with the graph, many characteristics of the plot can be adjusted. To obtain help on this command, enter

>   ?pointplot

To plot these points, enter

>   pointplot([P.(1..4)]);

Exercise 2

Enter the vectors $\mathbf{A} = \begin{bmatrix} 1 \\ 1 \end{bmatrix}$ and $\mathbf{B} = \begin{bmatrix} 2 \\ 3 \end{bmatrix}$.

>   A:=vector([1,1]); B:=vector([2,3]);

Using the **pointplot** command, plot these two vectors as points in the plane. Note the additional parameter *view*, which sets a viewing window for the graph.

>   pointplot([A,B],view=[0..3,0..3]);

a) Produce several different linear combinations of **A** and **B**. Appropriate syntax for doing this would be as follows.

>   evalm(2*A + 3*B);

b) Use Maple's random number generator to create several random linear combinations of A and B. First, define a function f which generates random four-digit numbers in the interval (-1,1). This is done with Maple's **rand** function as follows. The **evalf** command changes exact fractions to decimals.

>   f:=evalf(rand(-10000..10000)/10000):

Then, invoke the function f with the syntax f(). For example, create **c**, one random linear combination, with the syntax

>   c:=evalm(f()*A+f()*B);

1.2 Maple On Line

c) Plot enough points in the span of A and B to get a discernible geometric figure. Begin with 100 random linear combinations, and, using the **pointplot** command, plot them as points in the plane. Maple's **seq** command will produce a sequence of similar objects from a pattern provided to it. Terminate the command with a colon (:) to suppress the output. Then, feed the resulting sequence to the **pointplot** command, remembering to enclose the sequence in square brackets since **pointplot** requires a *list* of points (or vectors).

```
> q:=seq(evalm(f()*A+f()*B),k=1..100):
> pointplot([q]);
```

d) The plot in part (c) is only part of the span, the portion being determined by our use of random numbers in the interval (-1,1). Rather than redefine the function f, try multiplying the vector **A** by 2, creating another plot of at least 200 random linear combinations with **A** multiplied by 2.

Exercise 3

Describe in words the set of points corresponding to the collection of linear combinations defined by the sum $s\mathbf{A} + t\mathbf{B}$, where both s and t lie in closed intervals of the form [-2,2]. Plot the resulting set of linear combinations as points in the plane, using a different color than used in Exercise 2. (See the online help for how to specify color in the pointplot command.)

Exercise 4

In Exercise 9 of the non-computer problems, it was stated that each element $\begin{bmatrix} x \\ y \\ z \end{bmatrix}$ in the span of

$$\mathbf{X} = \begin{bmatrix} -1 \\ 1 \\ -1 \end{bmatrix} \text{ and } \mathbf{Y} = \begin{bmatrix} -1 \\ 3 \\ 2 \end{bmatrix}$$

satisfies $5x + 3y - 2z = 0$.

a) Write $\mathbf{C} = a\,\mathbf{X} + b\,\mathbf{Y}$, the expression for the general linear combination of A and Y. Then show that the components of the vector **C** satisfy the equation of the plane declared above. This is most effectively accomplished as follows.

```
> X := vector([-1,1,-1]); Y := vector([-1,3,2]);
> C := evalm(a*X + b*Y);
```

```
>  q := 5*x + 3*y - 2*z = 0;

>  q1 := subs(x=C[1], y=C[2], z=C[3], q);
```

b) Plot, as points in R^3, a few hundred elements of this span. This requires use of Maple's **pointplot3d** command, the 3d analog of **pointplot**. Begin by forming, via **seq**, a sequence of random vectors in the span of X and Y. The function f defined for Problem 1 can again be used to provide the random coefficients.

```
>  q:=seq(evalm(f()*X+f()*Y),k=1..200):
```

The syntax for **pointplot3d** is similar to that of **pointplot**. However, there are several additional parameters whose use makes for a better graph. Since all 3d plot packages attempt to plot a 3d object on a 2d sheet of paper (or computer screen), there is a need for a reference frame that provides the sense of depth. Try putting a box around your graph, either interactively via the toolbars, or via options to the plot command itself. Having clear and highly visible labels on the axes is equally useful. Thus, the following syntax could be used to plot the vectors randomly generated above.

```
>  pointplot3d([q], color=black, axes=boxed, labels=[x,y,z],
   labelfont=[TIMES,BOLD,14]);
```

Observe that in Maple, 3d graphs can be rotated on the screen by manipulation with the mouse. Click on the plot to make it "live." Then, click and hold down the mouse button, dragging the bounding box that now replaces the graph. This bounding box is rotated as the mouse is moved. When released, the bounding box has a new orientation, and the graph is redrawn by clicking the **R** (redraw) on the toolbar.

Try to rotate your graph to demonstrate that the plotted points lie on a plane.

1.3 On Line

Maple Release 4 permits more than one worksheet to be open at the same time. On some platforms Maple can be put into the "multiple kernels" mode in which each worksheet is attached to its own "kernel," or memory state. In this mode, variables declared in one worksheet will not be known to any other worksheet opened simultaneously.

However, the default setting for Maple might be the "shared kernel" mode in which all open worksheets share the same memory state. In this case, variables declared in one worksheet have the same value in every other worksheet opened simultaneously. The potential here for grave confusion

1.3 Maple On Line

is very high. Since, in Release 3, only one worksheet could be attached to a kernel, this conflict between multiple worksheets never arose. In Release 4 it is essential that this "gotcha" be understood. A simple precaution in the shared kernel world is starting every new worksheet with a **restart**, a Maple command which clears memory and resets all variables.

The use of the **restart** command at the beginning of each new worksheet is highly recommended.

Here, we both restart Maple and load the *linalg* package.

> `restart;`

> `with(linalg):`

Exercise 1

Obtain $\mathbf{X} = \begin{bmatrix} 4 \\ 0 \\ 0 \end{bmatrix} + s \begin{bmatrix} -2 \\ 1 \\ 0 \end{bmatrix} + t \begin{bmatrix} -3 \\ 0 \\ 1 \end{bmatrix}$ as the general solution to the equation $x + 2y + 3z = 4$.

Maple's **linsolve** command, the general linear system solver, will give the general solution to systems of equations. We begin by creating a matrix (the coefficient matrix) whose (i,j) entry is the coefficient of the j th variable in the i th equation. You can either type in the coefficient matrix A directly, or use the **genmatrix** command, giving a list of equations, and a list of variables as parameters. Thus, for the preceding linear equation, we could enter

> `q := x+2*y+3*z = 4;`

> `A := genmatrix([q],[x,y,z]);`

This extracts the coefficient matrix for the system. Next, the coefficient matrix and the constants on the right side of the equations are entered into **linsolve**. We express the constants as a list which, in this case, has only one entry.

> `X := linsolve(A,[4]);`

The arbitrary constants that Maple has introduced are $\_t_1$ and $\_t_2$. The lead character is the underscore (not a minus sign) and the numbers 1 and 2 are subscripts. To address these constants in Maple, use the syntax $\_t[1]$ and $\_t[2]$.

Setting these constants alternatively equal to 0 and 1 will "extract" the basis vectors in the general solution. After defining **u**, the translation vector, it is essential to subtract **u** from **X** when extracting the vectors **v** and **w**, the multipliers of $\_t_1$ and $\_t_2$.

```
> u := subs(_t[1]=0, _t[2]=0, op(X)); v := subs(_t[1]=1, _t[2]=0,
  evalm(X-u)); w := subs(_t[1]=0, _t[2]=1, evalm(X-u));
```

To get Maple to write the general solution in the vector form given in the statement of the problem, adroit use of the **evalm** command is necessary. Unless **evalm** is applied to the vector **v**, the screen will merely display the name, v.

```
> Xg := evalm(u) + s*evalm(v) + t*evalm(w);
```

a) With the vectors **v** and **w** declared as above, (or typed in afresh from the pencil-and-paper solution), form **C**, the general element in the span of **v** and **w**.

```
> C := evalm(a*v + b*w);
```

b) Show that $\mathbf{F} = \mathbf{u} + \mathbf{C}$ solves the given equation $x + 2y + 3z = 4$.

```
> F := evalm(u+C);
> q;
> subs(x=F[1], y=F[2], z=F[3], q);
```

This shows that the general solution of the given equation is the span of the vectors **v** and **w**, translated by the vector **u**.

c) Substitute the vector C into the given equation and describe the result. What should you conjecture from this result? Can you prove it?

Exercise 2

Repeat Exercise 1 for the system (U) from Section 1.3. The essential question to be ultimately resolved is "Does the conjecture made in Exercise 1 still hold?" Can you then prove your answer is correct?

Begin by entering the system (U), being careful to clear the variable w that was used in Exercise 1,

using the command " w:='w' ". Call your equations q1, q2, q3, and q4. Use **genmatrix** to form A, the coefficient matrix for the system (U).

```
> A := genmatrix([q.(1..4)],[x,y,z,w]);
```

Use **linsolve** to obtain the general solution of system (U), using a list for the values on the right hand sides of the equations.

```
> X := linsolve(A,[1,0,1,2]);
```

Extract the translation vector **u** and the basis vectors **v** and **w** as in Exercise 1.

a) Form **C**, the general element in the span of **v** and **w**.

1.4 Maple On Line

```
> C := evalm(a*v + b*w);
```

b) Show that **F** = **u** + **C** satisfies the system (U). This can be done by repetitive typing, by typing once and using copy/paste, or by a *for-loop* that does repetition automatically.

```
> F := evalm(u + C);
> for k from 1 to 4 do subs(x=F[1], y=F[2], z=F[3], w=F[4],
    q.k); od;
```

c) Substitute C into each equation in the system (U) in an attempt to determine if the conjecture made in Exercise 1 is still viable. If you still believe your conjecture is true, can you prove it?

1.4 On Line

After restarting Maple and reinitializing by loading the *linalg* package, we examine the **rref** command for putting a matrix into its reduced (row) echelon form.

Consider the matrix A defined by

```
> A := matrix(4,6, [1,-1,1,3,0,6,2,-2,2,6,0,7,-1,1,1,-1,-2,
    1,4,-4,1,9,3,6]);
```

Subject A to the **rref** operator.

```
> rref(A);
```

The result just obtained is exact since Maple obtained it by doing rational arithmetic. There is no truncation error introduced by a conversion of integers to decimal form, and there is no round-off error produced by a numerical algorithm. This is the result that would be obtained working with a pencil and paper.

This is, of course, wonderful. We can, in princple, do every computation with total accuracy. At first glance, then, it seems then we should never again need to round off an answer. Unfortuately, life is not so simple. Suppose, for example, we want MAPLE to compute 1.048577^{20}.

We enter

```
> (1048577/1000000)^20;
```

Maple's response is a 121 digit integer divided by an equally large power of 10. Imagine now, what would happen if we attempted to perform a calculation which required, say, addition and multiplication several hundred such numbers. The number of digits our computer would need to store would become astronomical and the speed would be reduced to a snail's pace. Furthermore, in an actual application, the number .1048577 would probably represent the result of some measurement which itself might be accurate only to within the given number of digits. Thus, the vast majority of the digits that our compter is so laboriously computing and saving are totally meaningless. The moral is that the perils of numerical computations must be faced.

Exercise 1

We first examine how to convert the matrix A to floating point (decimal) form, then look at the same row reduction done numerically instead of symbolically. The conversion can be done by *converting* each element to floating point form - **map** the convert operator onto the matrix A via

>   A1 := map(convert,A,float);

Row reduce to reduced echelon form the floating point form of matrix A.

>   rref(A1);

For the matrix A there is no difference in the reduced echelon form when working numerically. This will not always be the case. In fact, we can investigate Maple's numerics by a stratagem used on any numerical calculating device. Compute the value of $[\frac{1}{99}] 99 - 1$, and successively append 9's both inside and outside the brackets. Eventually, there will be enough 9's so that the numeric calculation will no longer yield 0. That gives an indication of how accurate the computing device is. To force Maple to evaluate the expressions in floating point form, make one of the numbers a decimal. For example, use "1.0" rather than just "1" in the numerator of the fraction.

To see the difference between working numerically and symbolically in Maple, change the numerator from "1.0" to just 1. Then, Maple will evaluate the expression symbolically and produce 0. The round-off error only appears when working with floating point numbers.

It is possible to vary the number of digits with which Maple computes. This is done via the Digits variable as follows.

>   Digits := 12;

Test Maple's numeric behavior on the floating point calculation above that failed to yield 0.

Notice that with more digits available, Maple escaped the effect of round-off in a computation that "failed" with just the default 10 digits.

1.4 Maple On Line

The price one pays for increasing the number of working digits is computation time, since these extra digits are being simulated by the Maple software.

Reset the number of digits back to the default 10 via

> Digits := 10;

Exercise 2

Use the **rref** command to find all solutions to the system in Exercise 5g, Section 1.3.

First, enter the equations of that system. Call your equations q1, q2, q3, qnd q4.

Next, get Maple to write the augmented system matrix. The **genmatrix** command converts the equations into matrix form, and the additional parameter *flag* signals Maple to include the numbers on the right side of the equations. Incidentally, the parameter can be any character or word that is not already a reserved word in Maple.

> A:=genmatrix([q.(1..4)],[x,y,z,w],flag);

Apply the **rref** command.

> A1:=rref(A);

To obtain solutions from the rref form of the matrix A, apply the process of back substitution. Start with the bottom-most non-zero row of rref(A) and interpret it as an equation defining the value of z. Solve that equation for the value of z and substitute that value into the equation above. Solve the resulting equation for the value of y so determined. Substitute both the value of y and z into the remaining equation which is then solved for x.

Check your work by invoking Maple's built-in **backsub** command.

> backsub(A1);

Exercise 3

The rank of a system of equations is the number of equations left after eliminating dependent equations. This number does not depend on which equations were kept or eliminated. Hence, it is plausible that the rows in the reduced row echelon form of the system's matrix reflect the distinct equations that would survive an elimination of dependent equations. Hence, the rank of the system should be the number of non-zero rows in the reduced row echelon form of the matrix for the system.

Check this conjecture experimentally by creating (4 x 5) matrices A1, A2, A3, and A4 with ranks respectively 1, 2, 3, 4. In particular, insure that no matrix has a zero entry.

A process for creating a random matrix of prescribed rank is based on forming rows that are themselves linear combinations of other rows. Begin by defining f, a function returning a random integer in the closed interval [-10,10]. This is done with the **rand** command.

>   f := rand(-10..10):

We begin by constructing a matrix A1, of rank 1. This requires that the rows of A1 be linear combinations of a single row. Begin by constructing a (1 x 5) matrix M1 by using the function f in conjunction with the **randmatrix** command to guarantee that the random matrix has entries that are integers in the interval [-10,10].

>   M1 := randmatrix(1,5,entries=f);

From M1, build a (2 x 5) matrix M2 in which the rows are multiples of the single row in M1. Maple's **stack** command assembles a rows (or vectors) into a new matrix, making the building-blocks into the rows of the new matrix.

>   M2 := stack(M1,evalm(f()*row(M1,1)));

From M2, build a (3 x 5) matrix M3 in which the rows are random linear combinations of the rows in M2. A single row in M2 can be referenced by the **row** command as illustrated below.

>   M3 := stack(M2,evalm(f()*row(M2,1)+f()*row(M2,2)));

Finally, build the required (4 x 5) matrix A1 by taking linear combinations of the rows of M3.

>   A1 := stack(M3,evalm(f()*row(M3,1) + f()*row(M3,2) + f()*row(M3,3)));

Test that A1 has rank 1 by invoking Maple's built-in **rank** command.

>   rank(A1);

The process for constructing random matrices of rank 2 is similar. The only difference is that we start with a random (2 x 5) matrix (produced using the **randmatrix** command) instead of a (1 x 5) matrix. Similarly, for a rank 3 matrix, we would begin with a random (3 x 5) matrix and for a rank 4 matrix we would begin with a random (4 x 5) matrix.

The rank of the matrices A1, A2, A3 and A4 can be corroborated by reducing each to reduced echelon form. The following loop implements the required calculations.

1.4 Maple On Line

```
> for k from 1 to 4 do rref(A.k); od;
```

In each case the number of distinct non-zero rows exactly matches the known rank of the matrix. These row reductions are exact, without round-off error since Maple computes symbolically unless told otherwise. However, all computing devices, when computing with floating point numbers, can experience difficulties attributable to round-off and truncation errors. Examine this issue in Maple.

```
> for k from 1 to 4 do rref(map(convert,A.k,float)); od;
```

For the matrices created in this session (remember, we are using a random process), rref(A3) is wrong. The **rref** command declares that a small number which ought to be seen as zero, is not zero. Hence, it suggests the rank of A3 is four.

One defense against such numeric errors is the **gausselim** command which row reduces a matrix but does not make the diagonal elements 1. By not dividing by the diagonal elements, this command is less likely to err in numeric computations. Let B3 be the floating point version of A3, obtained by mapping the process of conversion to floats onto A.

```
> B3 := map(convert,A3,float);
```

Now apply **gausselim**.

```
> q := gausselim(B3);
```

The small entries in the fourth row should be taken as 0's. These are the numbers that rref sees as non-zero, leading to errors. In Maple, we can apply the **fnormal** command to set to zero numbers smaller than a given tolerance. As with all operations applied to matrices and vectors, the **fnormal** command is mapped onto the matrix q.

```
> map(fnormal,q,10);
```

Exercise 4

Row reduce the transposes of the matrices A1, A2, A3, and A4 constructed in Exercise 3. The Maple command for the transpose is " transpose(A);" What do you notice about the rank of the resulting matrices?

Exercise 5

Define vectors **X**, **Y**, and **Z** as indicated below.

```
> X := vector([1,2,-5,4,3]); Y := vector([6,1,-8,2,10]); Z
:= vector([-5,12,-19,24,1]);
```

a) Determine which of the vectors **U** and **V** below is in the span of **X**, **Y**, and **Z**. Solve a determining system of equations by using the **rref** command.

> `U := vector([-5,23,-41,46,8]); V := vector([22,0,-22,0,34]);`

The question requires solving a **X** + b **Y** + c **Z** = **U**, and a **X** + b **Y** + c **Z** = **V** for constants a, b, and c. Both sets of equations can be solved at the same time if the following augmented matrix is formed.

> `q := augment(X,Y,Z,U,V);`

Row reducing via the rref command gives solutions to both systems of equations at the same time.

b) Imagine that you are the head of an engineering group and that you have a computer technician working for you who knows absolutely nothing about linear algebra, other than how to enter matrices and commands into Maple. You need to tell your technician how to do problems similar to part (a) above. Specifically, you will give the technician an initial set of three vectors, **X**, **Y**, and **Z** from R^3. You will then provide an additional vector **U** and you want the technician to determine whether **U** is in the span of **X**, **Y**, and **Z**.

Write a brief set of instructions which will tell your technician how to do this job. Be as explicit as possible. Remember that the technician cannot do linear algebra! You must provide instructions on how to construct the necessary matrices, what to do with them and how to interpret the answers. The final "output" to you should be a simple "Yes" or "No." You don't want to see matrices.

c) One of your assistant engineers comments that it would be easier for the technician in part (b) to use Maple's **rank** command rather than **rref**. What does your assistant have in mind?

1.5 On Line

After clearing Maple's memory by issuing a **restart** command, and re-initializing by loading the *linalg* package, enter the matrix A and the vector X.

> `A := matrix(3,4,[1,2,1,3,-5,7,2,2,13,4,4,3]);`

> `X := vector([1,3,-2,4]);`

Obtain the product **B** = A **X**. (You should consult On Line Section 1.1 for a discussion of matrix products in Maple.)

```
> B:=evalm(A&*X);
```

Exercise 1

The matrix multiplication A **X** just obtained represents a linear combination of the columns of A, with coefficients taken from the vector **X**. Implement this notion, and show the result is the vector **B** found in the Introduction.

Columns of A can be referenced with the **col** command, and elements of the vector **X** can be referenced as **X**[k]. Hence, the brute force way of obtaining the required linear combination would be with the following syntax.

```
> evalm(col(A,1)*X[1] + col(A,2)*X[2] + col(A,3)*X[3] + col(A,4)*X[4]);
```

Since the columns of A are referenced in numerical order with an index that is repeated when referencing the components of X, it should be possible to form the same linear combination of columns with some sort of summation process. Maple has a **sum** command that replicates exactly the mathematical sigma notation, $\sum_{k=1}^{4} A_k X_k$. There is one syntactical quirk to overcome, however. The **col** command requires a value for the index before the **sum** command can provide it, so naive use of the notation will result in a syntax error. The trick is to put single forward quotes on the **col** command, thereby preventing it from demanding priority in getting a value of the index before the **sum** command is ready to provide it.

```
> evalm(sum('col(A,k)'*X[k],k=1..4));
```

Exercise 2

Solve the system A **X** = **B** for **X**. Keep in mind that **B** was formed by multiplying A against **X**. This exercise seeks to determine whether or not **X** can be recovered from **B**.

One method of solution consists of row reducing the augmented matrix [A,B], then using back substitution, implemented in Maple via the **backsub** command.

```
> C := rref(augment(A,B));
> X1 := backsub(C);
```

By inspection, determine a value of the parameter $\_t_1$ that makes the general solution in X1 become precisely **X**. Remember, this parameter is a subscripted quantity, and can be addressed in Maple via the syntax _t[1].

Exercise 3

Another method for finding the general solution first obtained in Exercise 3 is predicated on finding a basis for the null space of A. This basis can be found via the Maple command **nullspace**, as shown below.

> `q := nullspace(A);`

Observe that the **nullspace** command returns a set of vectors. Here, there is but one member in the set, a single vector that can be addressed via the syntax

> `Z:=q[1];`

Verification that **Z** is indeed in the null space of A resides in the product A **Z**.

> `evalm(A &* Z);`

The general solution for the system A **X** = **B** is then **Xg** = **X** + t **Z**, where t is an arbitrary parameter. Form **Xg** and show that it satisfies the equation A **Xg** = **B**.

By inspection, determine a value of the parameter t in Xg for which Xg becomes exactly X1, the first form of the general solution found above.

Exercise 4

Enter the matrix A and the vector **B** as shown below.

> `A := matrix(4,6,[17,-6,13,27,64,19,4,-6,-33,25,7,9,55,-24,6,106,199,66,89,36,32,160,327,104]);`

> `B := vector([17,4,55,89]);`

a) Determine the rank of A. From this information, determine how many free variables the system A **X** = **0** will have.

b) How many spanning vectors will the null space of A contain?

c) Using Maple's **nullspace** command, find a spanning set for the null space of A. Since this command returns a set of vectors, extract all the vectors from this set, naming them w1, w2, etc.

> `q := nullspace(A);`

> `for k from 1 to 4 do w.k := q[k]; od;`

d) By inspection, find a vector **X** satisfying the equation A **X** = **B**. Verify that your guess indeed satisfies the equation.

e) If **F** is a general linear combination of the vectors **w1**, **w2**, ..., **w4**, show that **C** = **X** + **F** is still a solution to the equation A **X** = **B**. (Note:

1.5 Maple On Line

You might need Maple's **print** command as well as **evalm** to force Maple to display the results of your computations.)

f) Explain the statement "The general solution to $A\,X = B$ is the vector $X_o + W\,Y$, where W is the matrix whose columns are **w1, w2, w3**, and **w4**, and **Y** is any vector in R^4." Hint: Computationally, it would be useful to form the matrix W with the augment command, form the vector **Y** with four parameters for components, and to find the general solution to $A\,X = B$ via the linsolve command. This solution should match $X_o + W\,Y$.

g) Find a basis for the null space of A by solving the equation $A\,X = 0$ for the general solution, **X**. This is easily done in Maple by using the **linsolve** command. This command takes as arguments, the matrix A, and a vector (or list) of zeros as the right-hand side values. Maple will deliver a linear combination of the vectors **w1, w2, ..., w4** that were found by the **nullspace** command.

h) Find a basis for the null space of A. this time solving the system $A\,X = 0$ by augmenting A with a column of zeros and using the **rref** and **backsub** commands. The solution will not be readily recognized as a linear combination of the vectors **w1, w2, ..., w4**.

i) Verify that the basis found in part (h) is equivalent to the basis {**w1, w2, ..., w4**}, show that the set of equations $a\,\mathbf{w1} + b\,\mathbf{w2} + c\,\mathbf{w3} + d\,\mathbf{w4} = v_k$ has a solution for each v_k in the basis found in part (h). This is most easily done by augmenting W with the general solution found in part (h), and using the rref command to show the equations are consistent for any values of the parameters in that general solution.

Chapter 2

Dimension

2.1 On Line

Restart Maple to clear its memory of all variables, and re-initialize by loading the *linalg* package.

Exercise 1

Given the matrix A entered below,

```
> A := matrix(4,3,[1,2,-3,4,5,-1,3,2,1,1,1,1]);
```

find the reduced (row) echelon form of A. How can you tell just from this reduced form that the columns of A are independent? Relate your answer to Theorem 1.

Exercise 2

Let A be a (random) matrix with more rows than columns. State a general rule for using rref(A) to decide whether or not the columns of A are independent. Demonstrate your condition by (a) producing a 5 x 4 matrix A with no non-zero entries, and with independent columns; and by (b), producing a 5 x 4 matrix with no non-zero entries, and with dependent columns. In each case, obtain rref(A). Prove your condition using Theorem 1.

Exercise 3

Let A be the matrix entered below.

```
> A := matrix(5,6,[-1,2,6,-8,-14,3,2,4,1,-8,5,-1,-3,1,4,-9,
    -10,0,3,-2,-1,12,1,4,5,7,11,-11,-19,9]);
```

Use the **rref** command to find the pivot columns of A. Write them explicitly as columns. Then express the other columns of A as linear combinations of the pivot columns. (See Example 4 in the text.) You should discover that the first three columns of A are the pivot columns.

Exercise 4

If A_k represents the kth column of the matrix A defined in Exercise 3, form the matrix B whose columns are the columns of A in the following order: B = [A_4, A_6, A_1, A_2, A_3, A_5]. (This is most easily done by using the **augment** and **col** commands.) Find the pivot columns of B by using the **rref** command. Do you obtain a different set of pivot columns? Use rref(B) to express the other columns of B as linear combinations of the pivot columns. Could you have derived these expressions from those in Exercise 3? If so, how?

Exercise 5

Find a matrix C whose columns are just those of A listed in a different order, such that the column of C which equals A_5 and the column which equals A_1 are both pivot columns. Is it possible to find such a C where A_2 is a pivot column as well? If so, find an example. If not, explain why it is not possible.

2.2 On Line

Restart Maple to clear its memory of all variables, and re-initialize by loading the *linalg* package. In addition, use the command " with(student):" to load the *student* package in order to access its **equate** command.

Exercise 1

Let A be the matrix entered below.

```
> A := matrix(5,6,[-1,2,6,-8,-14,3,2,4,1,-8,5,-1,-3,1,4,-9,
    -10,0,3,-2,-1,12,1,4,5,7,11,-11,-19,9]);
```

2.2 Maple On Line

Part (a)

Find the rank of A via the **rank** command. Using only the value of the rank, explain why the statements *i*) and *ii*) given below are true.

i) The reduced form of the augmented matrix for the system A **X** = **0** has three free variables. (Recall that in a previous On Line section it was noted that the rank is the number of non-zero rows in the reduced form.)

ii) The null space of A has dimension 3. (Hint: How many spanning vectors are there in the general solution to A **X** = **0**?)

Part (b)

Show that each of the vectors **X1**, **X2**, and **X3** given below satisfy A **X** = **0**.

```
>   X1 := vector([-5,13,-10,2,-3,1]); X2 := vector([3,-6,11,1,2,-5]);
    X3 := vector([-4,7,9,5,1,-6]);
```

Note: Since verifying that A X_k = **0** is a repetitive task, it can be done in a *for-loop*.

Part (c)

By computing the rank of the matrix [**X1**, **X2**, **X3**], prove that **X1**, **X2**, and **X3** are linearly independent. (Recall that the maximal number of linearly independent columns equals the rank.)

Part (d)

How does it follow that the dimension of the null space of A is 3? How does it follow that the
X_k constitute a basis for the null space?

Part (e)

Using Maple's **nullspace** command, find a basis for the null space of A. Express each vector in this basis as a linear combination of the X_k's from part (d). Hint: Given two bases for this null space, showing that they are equivalent requires showing that any vector in one can be found as a linear combination of the vectors in the other. Hence, a set of equations of the form a X1 + b X2 + c X3 = w_k must be solved for each k = 1, 2, 3. This can be done by forming the augmented matrix [**X1, X2, X3, w1, w2, w3**] and row reducing. In row reduced form this matrix will indicate whether or not these equations are solvable, and if so, how to express the non-pivot vectors in terms of the pivot vectors. See Theorem 1 and Example 4 in Section 2.1.

Part (f)

In part (e), what made us so sure that the first three columns would be the pivot columns? Why couldn't, for example, the pivot columns be columns 1, 3, and 4? (Hint: Think about what this would imply for the reduced form of [**X1, X2, X3**].)

Part (g)

In part (e) you expressed the vectors w_k in terms of the vectors X_k. In this part, now express the X_k in terms of the w_k. This would complete the demonstration that the X_k and the w_k are equivalent spanning sets for the null space of A. Hint: Use the technique in part (e).

Part (h)

Find (by inspection) a vector **T** which solves the equation A **X** = $\begin{bmatrix} 6 \\ 1 \\ 4 \\ 1 \\ 11 \end{bmatrix}$.

Part (i)

Let $\begin{bmatrix} r \\ s \\ t \end{bmatrix}$ be an arbitrary element of R^3 and let **Z** = **T** + r **X1** + s **X2** + t **X3**, where **T** is as found in part (h). Compute A **Z**. Explain why you get what you get. Find constants u, v, and w such that **Z** = **T** + u **w1** + v **w2** + w **w3**. What theorem does this demonstrate?

Note: An efficient eay of doing this is to use the **equate** command from Maple. One might first enter the two proposed expressions for **Z** as follows:

> q1 := evalm(augment(X.(1..3))&*vector([r,s,t])); q2 := evalm
(augment(w.(1.

We can equate these two vectors with the **equate** command from the *student* package and then solve for the required constants using the following syntax.

> q3 := equate(q1,q2);
> q4 := solve(q3,{u,v,w}); q5 := solve(q3,{r,s,t});

2.3 On Line

Restart Maple to clear all variables and reset its memory, then initialize by loading the *linalg* package.

Exercise 1

Using Maple's **randmatrix** command, construct M, a random 3 x 5 matrix. What do you expect for the rank of M? Check, using the **rank** command. Is it conceivable that the rank could have turned out otherwise? Why is it unlikely? Finally, retain this matrix for use in the other exercises of this section.

Exercise 2

Form two different random linear combinations of the three rows of M, then append these two rows to M, thereby creating a 5 x 5 matrix. It helps to use **rand** to define a function f as a generator for the random coefficients needed for the linear combinations. Then, the **sum** and **row** commands simplify constructing the linear combinations of the rows of M. Finally, the **stack** command appends rows to the bottom of M. (See Exercise 3 from the On Line exercises for Section 1.4.) Using both **rank** and **rref**, test the rank of the enlarged matrix. What is the maximal number of linearly independent columns in the new matrix?

Exercise 3

For the 5 x 5 matrix created in Exercise 2, find a set of columns which forms a basis for the column space. Express the other columns as linear combinations of these columns. Use the technique of Example 3 in Section 2.1.

Exercise 4

In this exercise you will explore Maple's ability to obtain the reduced row echelon form numerically. In Exercise 3 the reduction is found symbolically, using exact arithmetic, and the result suffers no loss of accuracy from round-off error. A floating point evaluation of this exact answer will serve as the target answer that we will expect Maple to deliver numerically.

First, apply the **evalf** command to the reduced row echelon matrix found in Exercise 3. Next, by mapping the convert-to-float operation onto it, convert the 5 x 5 matrix of Exercise 2 to floating point form. (See

Exercise 1 in the On Line exercises for Section 1.4.) Then, obtain the reduced row echelon form of this numeric matrix. Observe that the result is wildly wrong. In fact, it shows the matrix to be of rank 4 whereas the matrix is known to have rank 3.

>   mm := map(convert,MM,float);

>   mm1:=rref(mm);

The reason for the error can be seen by row reducing the matrix to an upper-triangular form, without dividing the diagonal elements to make them 1's. This will prevent division by possible small numbers. This row reduction can be accomplished via the **gausselim** command.

>   q := gausselim(mm);

So, although the work was done numerically, the only difficulty that has surfaced is the small positive value on the main diagonal, a value that in exact arithmetic would be zero. To get Maple to render such small values as zeros, map the **fnormal** command onto the matrix. The **fnormal** command takes as additional parameter, an integer specifying the number of digits to which the rounding is to be performed.

>   rref(map(fnormal,q,9));

This result now matches what was obtained when the exact *solution* was converted to floating point form.

Exercise 5

Find a basis for the column space of the 5 x 5 of Exercise 2 by row reducing its transpose and invoking the Non-Zero Rows Theorem. Express each of the basis columns found in Exercise 3 as linear combinations of the basis columns found by this technique.

Exercise 6

Create four random 4 x 4 matrices of rank 1, 2, 3, and 4, respectively. (See Exercise 3 from the On Line exercises for Section 1.4.) Determine the null space and the dimension of the null space (called *nullity* in some texts) for each matrix. Relate the rank, the nullity, and the number of rows in the matrix.

Chapter 3

Transformations

3.1 Section 3.1 - On Line

Restart Maple to clear all variables, then load the *linalg* and *plots* packages. The exercises in this section deal with transformations in the plane. Hence, constructing several different types of graphs needed in this section requires the *plots* package.

Exercise 1

Create a 2 x n matrix F whose n columns are the coordinates of certain points in the plane. These points are the endpoints of line segments that constitute a letter of the alphabet. Because it might be tedious to construct the letter "O" with line segments, feel free to select some more easily constructed letter, such as "F".

Sketch the letter of your choice on a piece of paper, using a minimum number of arcs, and a maximum number of line segments. Pick an endpoint of some segment as the origin and assign coordinates to each of the other endpoints. Sketching the letter on a sheet of graph paper might make this easier.

Enter into the matrix F the coordinates of the endpoints of the segments making up your letter. Each column represents an endpoint. Start at an extremity, and use contiguous columns to represent points connected by a line segment. If your letter requires you to retrace a segment (this happens with the arms in letters like E and F), list the coordinates of any endpoints in the order in which they are traversed.

For example, a recognizable letter F can be represented by the matrix

$$\begin{bmatrix} 0 & 0 & 2 & 0 & 0 & \frac{3}{2} \\ 0 & 2 & 2 & 2 & 1 & 1 \end{bmatrix}$$

where the origin has been taken at the base of the vertical stroke.

Save the worksheet in which you have entered your matrix F since it will be used in the remaining exercises.

Exercise 2

Since you will need (and want?) to plot your letter, create the following Maple function that takes as input the name of the matrix of your letter, and returns a plot of the letter. The effort required to type in this function will more than pay for itself as you experiment with these exercises.

```
>   f := x -> pointplot([seq(convert(col(x,k),list),k=1..coldim(x))],
    style=line, axes=boxed, scaling=constrained):
```

Obtain a plot of your letter by applying the function f to the variable F associated with the data for your letter.

```
>   f(F);
```

Exercise 3

Example 1 from Section 3.1 contains a matrix M which represents a "shear along the x-axis." Call this shear matrix S_x and apply it to the letter stored in F by forming the matrix product
S_x F. Plot the image of the sheared letter.

Exercise 4

Construct the rotation matrix corresponding to a counterclockwise rotation of 20 degrees. Apply this rotation matrix to your letter and plot the result.

Since you will need other rotations, it will be more efficient if you build R, a "rotation matrix generating function" that accepts as input the number of degrees (counterclockwise) through which the rotation is to take place, and returns the matrix for this rotation. Maple's arrow notation for building functions is appropriate here. An appropriate syntax would be

```
>   R := x -> matrix(2,2,[cos(x),-sin(x),sin(x),cos(x)]);
>   R20 := R(20*Pi/180);
```

3.1 Maple On Line

Rotate your letter by multiplying it by R20 and plot the rotated letter by invoking the plotting function f built in Exercise 2.

Exercise 5

Create another letter, reduce it to a matrix of coordinates, and store it in an appropriate variable. For example, the letter E can be created by a simple modification of the matrix representing the letter F, and its matrix stored in the variable E. Plot the new letter. Then, in anticipation of plotting the combination F E, determine a way to plot the result of shifting E three units to the right. Call your shifted letter TE.

Hint: Since the translate of E three units to the right would have each x-coordinate increased by 3, you want to find a simple way to add 3 to each element in the first row of the matrix E. The Maple syntax 3$7 writes a sequence of seven 3's separated by commas. Hence, the following matrix has row of threes and a row of zeros.

```
> T := matrix(2,7,[3$7,0$7]);
```

Next, plot the combination FE. The utility function f which we built for graphing letters is not sophisticated enough to accept multiple inputs the way the standard plot functions in Maple will. Hence, create plots of each letter and combine the resulting graphics objects with the **display** command. Assign the plot of each letter to a variable, being sure to terminate each command with a colon. Then invoke **display**.

```
> p1:=f(TE):
> p2:=f(F):
> display([p1,p2]);
```

Exercise 6

Let S be the transformation of R^2 to itself wherein S(**X**) is a shift of **X** one unit to the right. Show graphically that S is not linear. Specifically, use the letter created in Exercise 1 to show that $S(2\,\mathbf{X}) \neq 2\,S(\mathbf{X})$.

Exercise 7

For each of the following, find a matrix M for which multiplication would accomplish the indicated transformation. In each case, validate your matrix by applying it to the letter created in Exercise 1.

Part (a)

Ma flips a letter upside down.

Part (b)

Mb flips a letter left-to-right.

Part (c)

Mc rotates a letter by 20 degrees counterclockwise

Part (d)

Md is a shear along y-axis.

Exercise 8

Plot the effect of each of the following transformations applied to the letter created in Exercise 1.

Part (a)

A shear along the x-axis followed by rotation of 20 degrees counterclockwise.

Part (b)

Rotate 20 degrees counterclockwise, then shear along the x-axis.

Part (c)

Shear along the x-axis, followed by shear along the y-axis.

Part (d)

Shear along the y-axis, followed by a shear along the x-axis.

3.2 On Line

Restart Maple to clear its memory of all variables, then re-initialize by loading both the *linalg* and *plots* packages.

These exercises will continue the study the geometric aspect of transformations in R^3. For this work it will be useful to again define the function f used in On Line Section 3.1. In that section the function took in a matrix representing a plane figure, and returned a plot of the figure.

Exercise 1

In place of the letters of the alphabet that were used in the exercises of Section 3.1, use line segments to create the outline of a car. Draw the outline on a piece of paper, even a sheet of graph paper, pick for the origin the endpoint of one segment, and find the coordinates of the endpoints of all the line segments. Enter these coordinates as columns of a matrix C. Hence, you might have a matrix such as the following.

```
>  C := matrix([[0,1,3/2,5,6,0],[0,1,5/2,2,0,0]]);
```

representing the car whose shape is given by the following graph.

```
>  f(C);
```

For the reader who believes this car looks more like a flat-iron, we give the data points for a car that is significantly better looking. The interested reader can enter the data into an appropriate matrix and run the experiments in these exercises with the improved image.

[[0, .954e-1], [.104e-1, .1612], [.1425, .2237], [.2124, .2336], [.2513, .2401], [.2902, .2401], [.3187, .3158], [.3472, .3553], [.3912, .3684], [.4611, .3750], [.5104, .3783], [.5674, .3783], [.6373, .3783], [.6710, .3684], [.6891, .3520], [.7047, .3388], [.7202, .3191], [.7306, .2961], [.7409, .2763], [.7876, .2664], [.8135, .2599], [.8264, .2500], [.8394, .2368], [.8472, .2072], [.8497, .1612], [.8497, .1513], [.8497, .1283], [.8497, .1020], [.26e-2, .1053]]

Transform your plane image into a 3d object by altering the matrix C as follows. Add a middle row of zeros by first adding a third row of zeros, then swapping the second and third rows. This can be done in Maple by first stacking C with a row of zeros, then using the **swaprow** command to interchange the new row of zeros with the original row 2.

```
>  d := swaprow(stack(C,[0$6]),2,3);
```

Since several 3d plots will be required, it is very useful to define a function f3 that will take in a matrix representing a 3d object, and return a 3d plot of the object so represented.

```
>  f3 := u-> pointplot3d([seq(convert(col(u,k),list),k=1..coldim(u))],
   style=line, axes=boxed, scaling=constrained, color=black, labels=[x,z,y],
   labelfont=[TIMES,BOLD,14]):
```

Having defined the function f3, apply it to the matrix d which represents a first version of a 3d car. Since this is a 3d plot, it can be rotated in Maple by clicking on the image and then using the mouse to "grab" and "rotate" the bounding box. Clicking on the R in the toolbar redraws the graph.

```
>  f3(d);
```

Exercise 2

Further dimension can be added to the image of the car by adding 1/4 to each element in row 3 of the matrix d, then augmenting the matrix d with the altered version e. One way to do this is to assemble (via stack) the first row of d, the altered second row, and then the third row.

```
>  e := stack(row(d,1),evalm(row(d,2)+1/4),row(d,3));
>  F := augment(d,e);
```

Test the efficacy of these improvements by plotting, using the function f3.

Exercise 3

Add some substance to the car by sketching in diagonal lines on each of the narrow faces. This requires alternating the columns of the augmented matrix F so that the first column comes from d, the second from matrix e, etc. In effect, build a new matrix FF by augmenting pairs of columns of the form [d_k e_k]. This is accomplished in Maple via the syntax

```
>  FF := augment(F,seq(op([col(d,k),col(e,k)]),k=1..coldim(C))):
```

The validation of the manipulation is in the plotting.

```
>  f3(FF);
```

Exercise 4

Some of the transformations that will be applied to the car include rotations. To keep the rotated car in the viewing window, it will help to move the origin to the center of the car. Deduce the coordinates of this center, and move the origin to that point by subtracting appropriate constants

3.3 Maple On Line

from the first and third rows of the matrix FF. Remember that the first row records x-coordinates, the second row, z-coordinates, and the third row, y-coordinates. Form this new matrix by altering the appropriate rows and reassembling them into a new matrix.

```
> FFF := stack(evalm(row(FF,1)-3),row(FF,2),evalm(row(FF,3)-3/2)):
```

Next, rotate the figure counterclockwise 30 degrees about the x-axis, then rotate that image 20 degrees counterclockwise about the z-axis. This is most easily done by building functions that yield the appropriate three-dimensional rotation matrices (Exercises 10, 11, and 12 in Section 3.1), then invoking the functions for the required angles. (See Exercise 4 in the On Line section for Section 3.1.)

Exercise 5

Obtain a single matrix whose action under multiplication reproduces the two successive rotations implemented in Exercise 4. Validate your single matrix by again plotting the rotated car.

Exercise 6

What image would you see if you transformed the matrix FFF by a rank 2 transformation? Create a random rank 2 matrix and test your guess. After printing graphs of the transformed and untransformed images, attempt to label several points where the transformation is many-to-one. The entries in the transformation matrix should be random numbers in the range [-1,1], lest the scale of the car be altered completely.

Exercise 7

What image would you see if you transformed the matrix FFF by a rank 1 transformation? Create a random rank 1 matrix and test your guess. Again, be sure to restrict the entries in your random matrix to the range [-1,1].

3.3 On Line

Restart Maple to clear its memory of all variables, then reinitialize by loading the *linalg*, *plots*, and *plottools* packages.

Exercise 1

Create M, a random 2 x 3 matrix with rank 1. (See Exercise 3 in Section 1.4.) If M is interpreted as the matrix of a transformation acting on R^3, what should the dimension of the image of this transformation be? Verify this by creating 100 random points in R^3, transforming them under M, and plotting the points.

Next, generate P, a matrix containing 100 random points in R^3. A moment's thought about how the transformed points will be plotted will determine the optimum strategy for generating and plotting the points. If the points are stored as columns of a matrix, they can be plotted by applying the **pointplot** command to the matrix. Hence, let P be of dimension 3 x 100 so the product M P will be 2 x 100. Create P by juxtaposing, via **augment**, 100 vectors generated by **randvector**. Terminating the commands with a colon (:) signals Maple not to print the rather large outputs to the screen. Finally, anticipating Exercise 2 where this plot will be required, store the plot data structure in a variable, say p1, so later, other images can be superimposed on it.

```
>  P := augment(seq(randvector(3),k=1..100)):

>  S := evalm(M&*P):

>  p1 := pointplot(S):

>  p1;
```

Exercise 2

The plot generated in Exercise 1 should show the span of any non-zero column of M. Demonstrate this by choosing a column of M and plotting, on the graph from Exercise 1, 100 random points in the span of this column.

Define the function f which generates a random integer in the closed interval [-500,500]. Then, using the **seq** command, give to **pointplot** a sequence of 100 random multiples of the first column of M. Color the points red and assign the plot data structure to a variable, say p2. After viewing the graph, merge it with the plot from Exercise 1 by use of the **display** command. Assign this composite graph to a variable, say p3, for use in Exercise 3.

```
>  f := rand(-500..500):

>  p2 := pointplot([seq(evalm(col(M,1)*f()),k=1..100)], color
 = red):   p2;

>  p3 := display([p1,p2]):   p3;
```

3.3 Maple On Line

Exercise 3

Using information and insights from Exercises 1 and 2, find (a) a specific vector **B** in R^2 for which the equation M **X** = **B** is *not* solvable; and (b) a vector **C** in R^2 for which the equation M **X** = **C** *is* solvable. Indicate these vectors on the composite graph produced in Exercise 2. Verify your answers by computing, via **rref**, the reduced row echelon forms of the augmented matrices [M,**B**] and [M,**C**].

Exercise 4

Plot 100 random elements from the null space of M. Maple's nullspace command will provide a basis for the null space of M. This basis is returned as a set of vectors which can be extracted from the set with the bracket notation. The **seq** command can be used to generate a sequence of 100 random linear combinations of these basis vectors, a sequence which can then be plotted in R^3 with the **pointplot3d** command. The 3d graph so generated can be rotated on-screen by grabbing and rotating the bounding box. It should then be possible to observe the nature of that portion of the span so generated.

Finally, explain how this plot relates to the Rank-Nullity Theorem.

Exercise 5

Randomly generate a rank 2 matrix M of dimension 3 x 3. Repeat Exercises 1 through 4, suitably modified to account for the different dimensions.

Specifically, this means you are to generate M. Then, in imitation of Exercise 1, the matrix P containing 100 random point in R^3, and plot the product M P. In imitation of Exercise 2, plot 100 random linear combinations of the columns of M. In imitation of Exercise 3, find vectors **B** and **C** for which the systems M **X** = **B**, and M **X** = **C** are not solvable, and solvable, respectively. Verify your choices of **B** and **C** computationally. Finally, in imitation of Exercise 4, plot 100 randomly chosen elements from the null space of M.

When you are done, don't forget to relate your findings to the Rank-Nullity Theorem.

3.4 On Line

Restart Maple to clear its memory of all variables, and re-initialize by loading the *linalg* and *student* packages.

Maple contains a variety of "solvers" for equations of various types. For example, the standard symbolic solver for one or several equations, linear and non-linear alike, is **solve**. The standard numeric solver for such equations would be **fsolve** (floating point solve). Differential equations are solved by **dsolve**, difference equations are solved by **rsolve** (recursive solve) and Diophantine equations are solve by **isolve** (integer solve).

In the linalg package, if a set of linear equations is captured in the matrix-vector format A **X** = **B**, then **linsolve** can be used. Both **solve** and **linsolve** return general symbolic solutions when applicable. In fact, **linsolve** will even return solutions in terms of arbitrary parameters. Other approaches to the solution of linear systems include the use of **gausselim** (followed by **backsub**) or **rref** (also followed by **backsub**). Maple's command structure is rich enough that nearly any undergraduate mathematics that can be articulated in standard mathematical notation can probably be implemented in the context of Maple's built-in commands.

There are two cautions to observe when using Maple to solve linear systems. First, if the calculation is done in floating point arithmetic, Maple is as liable to round-off and truncation errors as any other numeric utility. Second, when working symbolically, exact expressions for numbers can get dauntingly large, thereby consuming memory and time. Hence, Maple cannot solve symbolically systems as large as some strictly numeric utilities can solve by working in floats.

Exercise 1

Let A be the matrix from Exercise 2(a) of Section 3.4. Solve the system A **X** = **B** where

$$\mathbf{B} = [\tfrac{1}{10}] \begin{bmatrix} 21 \\ 32 \\ -44 \end{bmatrix},$$ then convert the answer to floating point form.

Next, convert both A and **B** to floats by mapping the convert/float operator onto them. Re-solve the system and compare the two floating point results. (See Exercise 4 in Section 2.3.)

Exercise 2

In many applications of linear algebra, numerical data comes from measurements which are susceptible to error. Suppose the vector **B** in Exercise 1 was obtained by measuring a vector **Ba** whose actual value is **Ba** = $[\frac{1}{100}] \begin{bmatrix} 210 \\ 321 \\ -440 \end{bmatrix}$. Compute the solution to the equation A **Xa** = **Ba**.

Measure error is the absolute value of the difference between the computed value and the actual value. It can be computed in Maple with the following command.

```
> e := map(abs,evalm(X-Xa));
```

Which component of the solution **X** computed in Exercise 1 has the largest error? (It might hepl to convert your answer to floting point form.) In terms of the magnitude of the components of the inverse $A^{(-1)}$, explain why this is the largest to be expected. (Maple computes the inverse of a matrix with the **inverse** command.) Which component of **Ba** would you change in order to produce the greatest change in **Xa**? Why? Back up your answer with a numerical example or with a symbolic calculation where the increments in **Ba** are parameters successively appearing in each component. How much error could you tolerate in the measured values of the components of **B** if the absolute value of the error in any entry of **X** is to be at most .001?

Exercise 3

Let A and **B** be as defined below.

```
> A := matrix(3,3,[1,1/2,1/3,1/2,1/3,1/4,1/3,1/4,1/5]);
> B := vector([83,46,32]);
```

Find the solution to A **X** = **B**. As in Exercise 2, suppose the vector **B** was obtained by measuring a vector **Ba** whose actual value is **Ba** = $[\frac{1}{100}] \begin{bmatrix} 8290 \\ 4607 \\ 3130 \end{bmatrix}$. Solve the equation A **Xa** = **Ba**. What is the percentage error in the least accurate entry of **X**? How much error could you tolerate in the measured values of the components of **B** if the absolute value of the error in any entry of **X** is to be at most .001?

Exercise 4

Exercises 2 and 3 demonstrate that the process of solving a system of equations can "magnify" errors in disastrous ways. One quantitative measure of the inaccuracy of a calculation is the ratio of the percentage error in the final answer to the percentage error in the input data. But what do we mean by the percentage error in a vector (such as **X** in Exercises 1 and 2) in which every component might have errors of different magnitudes?

For vectors in R^3, this question has a geometric meaning. Think of X_1 and X_2 as representing points in R^3. The distance d between these points is one measure of the error. If $X_1 = [x_1, y_1, z_1]^t$ and $X_2 = [x_2, y_2, z_2]^t$, then

$$d = \sqrt{(x_1 - x_2)^2 + (y_1 - y_2)^2 + (z_1 - z_2)^2}.$$

In Maple, this can be computed as "norm($X_1 - X_2$, 2)". The additional "2" represents the "2-norm" wherein differences are squared and the square root of the sum taken. If X_1 is the computed answer and X_2 is the actual answer, we define the percentage error to be

$$P = 100 \text{ norm}(X_1 - X_2, 2)/\text{norm}(X_1, 2).$$

a) Let **B**, **Ba**, **X** and **Xa** be as in Exercise 2. Use the given formula to compute (i) the percentage error in **B**, (ii) the percentage error in **X**, and (iii) the ratio of the percentage error in **X** to that in **B**. This is the inaccuracy of the calculation of **X** from **B**. Assuming that accuracy is desired, do we want the number d to be large or small? Explain.

b) Compute the inaccuracy of the computation of **X** from **B** in Exercise 3.

c) For each n x n invertible matrix A there is a number "cond(A)" (the "condition number of A") such that the inaccuracy in solving the system A **X** = **B** is at most cond(A), regardless of **B** and regardless of the amount of error in **B**. This means that if, say, cond(A) = 20 and the error in **B** is .001%, then the computed value of **X** will have at most 20 x .001 = .02% error. In general, $1 \leq$ cond(A). (This says that we cannot expect the answer to be more accurate than the input data.)

Maple has a built-in command for the condition number. Since the condition number is constructed from a "norm" of the matrix A, we need to specify the version of the condition number we want, according to the type of norm we want used. Hence, we will use the syntax cond(A,2). In addition, the 2-norm of a symbolic matrix can be a very large and complex expression. For this reason, we will only compute the condition number, based on the 2-norm, of matrices of floating point numbers.

3.5 Maple On Line

Compute the condition numbers for the coeficient matrices in Exercise 1 and 3. Use this to explain the difference in accuracies obaine in these exercises.

Matrices with large condition numbers are called "ill-conditioned." If the coefficient matrix of a system is ill-conditioned, then we must be extremely suspicious of answers obtained by solving the system since any slight error in the input data can make the solution very inaccurate. Notice that these inaccuracies are not related to round-off error. Ill-conditioning is intrinsic in the matrix and not in the method of solution.

3.5 On Line

Restart Maple, clearing its memory of all defined variables. Then, re-initialize by loading the *linalg* and *student* packages.

These exercises will explore **LUdecomp**, Maple's built-in command for obtaining the LU decomposition of a matrix. The LUdecomp command can return seven different items, namely, L, U, a factorization of U into U1 and R, a permutation matrix P, the determinant of U1, and the rank of A. We will not need all of these outputs, and will restrict explorations to just L, U, and P.

The syntax for a multi-return Maple function is tedious. Each variable that is to have a return assigned to it must be included in the command surrounded by single quotes.

Since the actual return of the **LUdecomp** command is U, we will not need to make a specific request for U to be returned. Thus, to obtain L, U, and any permutation P needed to complete the decomposition, the appropriate syntax would be

$$u := \text{LUdecomp}(A, L = 'l', P = 'p');$$

Of course, if only L and U were desired, then the command could be shortened to

$$u := \text{LUdecomp}(A, L = 'l');$$

Exercise 1

Enter the Matrix A from Example 1 of the text, then obtain the LU decomposition by using the **LUdecomp** command. Assign the output of this command to the variable u, and let l be the lower triangular factor. This matrix will not need a permutation P, so it can be omitted from the command. Finally, verify that A = l u. (You may need to use either **print** or **evalm** to view the contents of l.)

Exercise 2

Before examining the consequences pivoting has on the LU decomposition, we study the notion of a matrix factorization. When we demand that A be "factored" into the product LU, with L being a lower-triangular, and U being an upper-triangular matrix, we are asking for the solution of a set of equations. For example, if A is a given 3 x 3 matrix, then finding matrices L and U such that A=LU is equivalent with solving a system of nine equations, where each equations is obtained by setting one entry of A equal to the corresponding entry of LU.

In this exercise, we investigate this system for the matrix A of Exercise 1. Begin by forming L and U, 3x3 matrices of indeterminates L_{ij} and U_{ij}.

Maple's **matrix** command, with appropriate if-statements as an option, will create the desired matrices.

```
> L := matrix(3,3,(i,j)-> if i<j then 0 else L.i.j; fi); U
:=matrix(3,3,(i,j)-> if i>j then 0 else U.i.j;fi);
```

Multiply L and U, forming a template of indeterminates which we then demand reduce to the entries of A from Exercise 1, above. This gives a set of nine equations in twelve unknowns, since there are six unknowns in each of L and U. The solution of this set of equations will not be unique, suggesting that we can impose an additional three conditions on the factorization.

```
> LU := evalm(L &* U);
```

The product LU and the matrix A can be equated via the **equate** command from the *student* package. This command returns a set of equations formed by equating corresponding entries of each matrix.

```
> q := equate(LU,A);
```

Typically, the **solve** command needs a set of equations and a set of variables. If no variables are suggested to Maple, it will attempt to deduce what the unknowns of the problem actually are. That is convenient here since it would be tedious to enter the names of all the variables in these equations.

```
> q1 := solve(q);
```

It is clear that we did not get a unique solution for the entries of L and U. Careful inspection shows there are three indeterminates in the answer.

3.5 Maple On Line

This becomes more evident if we substitute these solutions into the matrices L and U.

```
> L1 := subs(q1,op(L)); U1 := subs(q1,op(U));
```

Exercise 3

Change the definition of the matrix L used in Exercise 2. Since there are three free parameters in the solution for the factors L and U, choose to have the diagonal elements of L all be 1's. This can be done with an appropriate if-statement in the **matrix** command that defines L. The matrix U will be the same as used in Exercise 2. Repeat the formation of nine equations in nine unknowns, obtaining unique factors L and U for the matrix A in Exercise 1. Display the resulting matrices L and U, and show that they are exactly the factors produced by the **LUdecomp** command in Exercise 1.

Exercise 4

In this exercise you will explore the concept of a permutation matrix P whose rows are a permutation of the rows of the identity matrix. If a matrix A is multiplied by P, the rows of PA will be permuted in the same way that the rows the identity were when forming P. Let P be a permutation of the rows of the 4x4 identity. This can be done in Maple by stacking a sequence of rows from the identity. Create A, a random 4x4 matrix and examine A, PA, and P.

```
> Id := diag(1$4);
> P := stack(seq(row(Id,k), k = [2,4,3,1]));
> A := randmatrix(4,4);
> PA := evalm(P &* A);
> print(A,PA,P);
```

Finally, observe that for a permutation matrix P, the inverse is the transpose.

```
> print(inverse(P),transpose(P));
```

Exercise 5

We wish to study the effect of pivoting on the LU decomposition. For this, we need a matrix A that forces a pivot. If the (1,1)-element of A were to be zero, a pivot would have to be performed immediately, since the row-reduction process by which we obtain L and U is basically gaussian elimination. Create A, a random 4x4 matrix, and then re-assign its (1,1)-element the value zero.

Use the **LUdecomp** command to obtain the LU decomposition, this time including parameter P = 'p' so that a permutation matrix p will be returned. To have Maple display p, l, and u side-by-side, use the **print** command.

```
>   u := LUdecomp(A,L='l',P='p'):   print(p,l,u);
```

Use the **print** command to display, side-by-side, the product lu, the matrix A, and the product plu.

You should observe that the product lu does not reproduce A. That is the effect of pivoting. Whenever rows must be interchanged during the factorization, these interchanges are recorded in the matrix P. The resulting LU factorization is then a factorization of $P^{(-1)}$ A, not A, resulting in the equality $P^{(-1)}$ A = LU. Thus, A = PLU, not just LU.

Chapter 4

Orthogonality

4.1 On Line

Restart Maple to clear its memory of all defined variables, and re-initialize by loading the *linalg, plots, student,* and *plottools* packages.

Given the vectors $Q_1 = \begin{bmatrix} -1 \\ 1 \end{bmatrix}$ and $Q_2 = \begin{bmatrix} 1 \\ 2 \end{bmatrix}$, we will generate a coordinate grid corresponding to a space in which Q_1 and Q_2 are the basis vectors. We will plot the skewed grid lines in red, atop a standard coordinate system in black.

```
> Q1 := vector([-1,1]); Q2 := vector([1,2]);
```

The vector equation of a line through the tip of the vector Q_1 and parallel to the vector Q_2 is given by $\mathbf{R}(t) = Q_1 + t\, Q_2$. To form parallel lines through the tip of 2 Q_1, 3 Q_1, etc., we need to form the vectors k Q_1 + t Q_2, with k an integer in some interval [-a,a]. Alternatively, the equations of lines parallel Q_1 through the tip of Q_2, 2 Q_2, etc, we for the vectors k Q_2 + t Q_1. In every case we let t, the parameter of the line, range over an interval [-b,b].

Maple's seq command can be used to generate an appropriate sequence of vector representations of the grid lines. Plotting them in red will product the desired grid.

A template for the vector form of each set of grid lines is obtained as

```
> P1 := evalm(k*Q1+t*Q2); P2 := evalm(k*Q2+t*Q1);
```

Sequences of equations of the skew grid lines are formed with the **seq** command.

```
> s1 := seq([P1[1],P1[2],t=-5..5],k=-5..5):   s2 := seq([P2[1],
P2[2],t=-5..5]
```

The plot of the grid lines is assigned to the variable f1 so that it can be used again in one more activity. We include the scaling parameter in the **plot** command. A 1-1 scaling can also be imposed interactively from the toolbar. A view window is also set in the **plot** command.

```
> f1 := plot({s1,s2}, color=red, scaling=constrained, view=
[-6..6,-6..6]):f1;
```

For a finishing touch, we use the **arrow** command from the *plottools* package to draw Q_1 and Q_2, the basis vectors of the skew coordinate system. We plot the first in green and the second in blue, superimposing both on the skewed red grid.

```
> a1 := arrow([0,0],[1,2], .2,.4,.2,color=green):   a2 := arrow
([0,0],[-1,1], display([f1,a1,a2]);
```

Exercise 1

Modify the commands in the Introduction to produce red grid lines corresponding to the basis vectors $Q_1 = \begin{bmatrix} 1 \\ 2 \end{bmatrix}$ and $Q_2 = \begin{bmatrix} 1 \\ -2 \end{bmatrix}$. In addition, produce a graph showing the skewed grid lines and the basis vectors in green and blue. Assign this graph to a variable so that it can be re-used in Exercise 4.

Exercise 2

The curve defined implicitly by the equation $x^2 - \frac{y^2}{4} = 1$ is a hyperbola that will be plotted in Exercise 3. Here, show that in the basis of Exercise 1 (with coordinates u and v) the equation of this hyperbola is $4uv = 1$.

Begin by deducing the equations of the transformation. The point matrix M is constructed with columns Q_1 and Q_2. Letting $\mathbf{X} = \begin{bmatrix} x \\ y \end{bmatrix}$ and $\mathbf{U} = \begin{bmatrix} u \\ v \end{bmatrix}$, the transformation equations are $\mathbf{X} = \mathbf{M}\,\mathbf{U}$.

The vectors **X** and M U can be equated with the **equate** command from the student package. The equation $\mathbf{X} = \mathbf{M}\,\mathbf{U}$ defines the change of basis from xy-coordinates to uv-coordinates. Once these equations are obtained, Maple's **subs** command can be used to impose this change of coordinates on the equation of the hyperbola. An appropriate syntax might be

```
> q1 := equate(X, MU);
```

4.2 Maple On Line

```
> q2 := x^2 - y^2/4 = 1;
> q3 := subs(q1,q2);
> q4 := simplify(q3);
```

Exercise 3

Use Maple's **implicitplot** command from the *plots* package to obtain a graph, in the xy-coordinates system, of the original hyperbola. Assign the graph to a variable so that in Exercise 4 it can be re-used. Be sure to use 1-1 scaling so that no distortion is introduced by the computer screen.

```
> g3 := implicitplot(q2,x=-10..10,y=-10..10,color=black,
  scaling=constrained): g3;
```

Exercise 4

Use Maple's **display** command to superimpose on the skewed grid lines of Exercise 1, the graph of the hyperbola from Exercise 3. Then use Maple's digitizer (click on the graph, click on a point in the graph, read the coordinates in the window at the top-left of the graphics toolbar) to approximate the coordinates of some point on the hyperbola. By counting and estimating, infer the corresponding uv-coordinates. Show that to within the accuracy of the digitizer, your coordinates satisfy the equation of the hyperbola in the *xy*-system and in the *uv*-system.

For example, the point (2,4) appears to be almost on the hyperbola. The digitizer gives (2, 3.5). The same point appears to be (2, 0.1) in the skew grid. Hence,

```
> subs(x=2,y=3.5,q2);
> subs(u=2,v=.1,q4);
```

4.2 On Line

Restart Maple, clearing its memory of all defined variables. Then, re-initialize by loading the *linalg* and *student* packages.

You are working for an engineering firm and your boss insists that you find one single solution to the following system:

$$2x + 3y + 4z + 3w = 12.9$$

$$4x + 7y - 6z - 8w = -7.1$$
$$6x + 10y - 2z - 5w = 5.9$$

You object, noting that

(a) The system is clearly inconsistent: the sum of the first two equations contradicts the third.

(b) You need at least four equations to determine four unknowns uniquely. Even if the system were solvable, you couldn't produce just one solution.

The boss won't take "no" for an answer. Concerning (a), the boss points out that the system was obtained from measured data and any inconsistencies can only be due to experimental error. Indeed, if any one of the constants on the right sides of the equations were modified by .1 units in the appropriate direction, the system would be consistent.

Concerning (b), the boss says "Do the best you can. We will pass this data on to our customers and they wouldn't know what to do with multiple answers."

After some thought, you realize that projections can help with the inconsistency problem The given system can be written in vector format as

$$x \begin{bmatrix} 2 \\ 4 \\ 6 \end{bmatrix} + y \begin{bmatrix} 3 \\ 7 \\ 10 \end{bmatrix} + z \begin{bmatrix} 4 \\ -6 \\ -2 \end{bmatrix} + w \begin{bmatrix} 3 \\ -8 \\ -5 \end{bmatrix} = \begin{bmatrix} 12.9 \\ -7.1 \\ 5.9 \end{bmatrix}$$

You realize that this system would be solvable if the vector on the right side of the above equation were in the space spanned by the four vectors on the left. Using Maple's rank command, you quickly compute as 2, the rank of the system matrix A,

$$A = \begin{bmatrix} 2 & 3 & 4 & 3 \\ 4 & 7 & -6 & -8 \\ 6 & 10 & -2 & -5 \end{bmatrix}$$

showing that these four vectors in fact span a plane. Call this plane W.

Your idea is to let **Bw** be the projection of $\mathbf{B} = \begin{bmatrix} 12.9 \\ -7.1 \\ 5.9 \end{bmatrix}$ onto W. Since the system is so nearly consistent, **Bw** should be very close to **B**. Furthermore, the system A **X** = **Bw** should certainly be solvable and one of the solutions should be what the boss is looking for.

Your point (b) will require some further thought. However, you do eventually come up with an idea which will be described in the exercises which follow.

4.2 Maple On Line

Before going on to the exercises, it will be useful to enter the data of this system of equations. Enter the matrix A and the vector **B**. Convert **B** to exact rational form and call that vector **b**. Compute the rank of A, verifying that it is indeed 2.

```
>   A := matrix(3,4,[2,3,4,3,4,7,-6,-8,6,10,-2,-5]);
>   B := vector([12.9,-7.1,5.9]);
>   b := map(convert,B,rational);
>   rank(A);
```

Exercise 1

Find an orthogonal basis for the column space of A, use the Fourier Theorem to obtain **Bw**, the projection of **B** onto W, the column space of A. Then solve A **X** = **Bw**, expressing the solution in parametric form. Write **X** as a sum of a "translation" vector, and vectors in the null space of A. Show that the translation vector Maple finds is not orthogonal to the null space of A.

First, obtain an orthonormal basis for the span of the column space of A. Since the columns of A are not linearly independent (A has rank 2), determine (via **rref**) which two columns of A to take as independent, and pass those two vectors to Maple's **GramSchmidt** command for orthogonalization. This command returns a list of *lists* (not vectors), so map the convert-to-vector operator onto this output. You will now have a list of two orthogonal vectors. An appropriate syntax might be:

```
>   q := map(convert,GramSchmidt([col(A,1),col(A,2)]), vector);
```

Next, apply the Fourier Theorem to get **Bw**, the projection of **B** onto W, the column space of A. Work with **b**, the exact version of the vector **B**. The formula in the Fourier Theorem is implemented in Maple much as it is written mathematically. Reference the basis vectors as q[1] and q[2], compute dot products with Maple's **dotprod** command, and norms with Maple's **norm** command, being sure to compute the 2-norm.

Next, solve the system A **X** = **Bw**. Maple's **linsolve** command will yield the general solution.

This general solution can be put into the form $X = T + \alpha\, v1 + \beta\, v2$, where the vectors **v1** and **v2** are in the null space of A. These vectors can be extracted from this general solution by an adroit use of substitution via the **subs** command.

Exercise 2

Objection (b) amounts to this: If the system is inconsistent, there is no solution. If the system is consistent, there are many solutions. Even the use of projections in Exercise 1 has yielded a general solution that is not unique. Perhaps your first thought was to report the translation vector as the solution. But there is nothing special about this vector, and the Translation Theorem says the general solution can be expressed using any particular solution, not just the translation vector.

Your next idea, however, is sound. With **T** as the translation vector, let **Tn** be its projection onto the null space of A. Report to your boss the solution **X** = **T** - **Tn**. Why is this **X** a solution?

Use Maple's **nullspace** command to find a basis for the null space of A. Then use the Fourier Theorem to obtain the projection of **T** onto this null space. Finally, form **T** - **Tn** and explain why this is a solution. Note, however that the Fourier Theorem assumes an orthogonal basis for the space into which the projection occurs. Hence, you will need to apply Maple's **GramSchmidt** command to the basis for the nullspace. Remember, though, that **GramSchmidt** returns a list of lists, requiring us to convert the sub-lists back to vectors an was done in Exercise 1.

Exercise 3

Try computing **X** in Exercise 2 by starting with a solution other than **T**. You should get the same **X**. Why? It can be shown that the **X** found in Exercise 2 is the solution of minimal length.

Exercise 4

Show that of all solutions to A **X** = **Bw**, the one with minimal length is the solution **X** computed in Exercise 3. For this, let **q1** be the general solution found by Maple in Exercise 1, considered as a two-parameter family of vectors. Compute the 2-norm, and use calculus to minimize this norm. The resulting solution should be the same **X** that was computed in Exercise 3.

Hint: Use the **subs** command to repace the free parameters _t[1] and _t[2] used by Maple with a and b, calling the result **Xg**. Now, obtain the 2-norm of the vector **Xg**. Since **Xg** is a symbolic vector, Maple returns the norm with absolute values, thereby making it difficult to differentiate and set derivatives equal to zero. Simplify the norm of **Xg**, adding the parameter *symbolic* to coax Maple to simplify the absolute values.

4.3 Maple On Line

```
> f := simplify(norm(Xg,2),symbolic);
```

Differentiate with respect to a and b. Use Maple's **solve** command to solve the system obtained by setting these partials equal to zero, as follows. Finally, substitute these values into **Xg** and cmpare with **X**.

```
> fa := diff(f,a); fb := diff(f,b);
> qq := solve({fa,fb},{a,b});
```

Exercise 5

Maple would have found the least squares solution of minimal norm with its built-in **leastsqrs** command. Verify that this command yields the solution X found in Exercise 2. Note the inclusion of the parameter *optimize* which signals Maple to find the solution of minimal length. Without this parameter, the **leastsqrs** command will return the general solution found in Exercise 1 as the vector in q1.

```
> leastsqrs(A,b,optimize);
> leastsqrs(A,b);
```

Exercise 6

After giving the boss your answer, you delete all your data except for A and **X**. A month later the customer calls, saying, "We know that there must be other solutions. Could you please provide us with the general solution?"

Show how the general solution can be reconstructed from A and **X** with a single Maple command.

4.3 On Line

Restart Maple to clear its memory of all previously defined variables. Then, re-initialize by loading the *linalg* and *plots* packages. In addition, enter the following lines of Maple code. This code, written by Dr. Mike Monagan of Simon Frasier University in Burnaby, British Colombia, Canada, creates a function that will generate the periodic extension of a function. The code first appeared in the article <u>Tips for Maple Users and Programmers</u>, *MapleTech*, VOL. 3, NO. 3, 1996, published by *Birkhauser*.

PE := proc(f, d::range) subs({'F' = f, 'L' = lhs(d), 'D' = rhs(d)-lhs(d)}, proc(x::algebraic) local y; y := floor((x-L)/D); F(x-y*D); end) end:

These lines of code can be entered into a separate Maple worksheet, and that worksheet saved. Later, if the code is again needed, that worksheet can be opened, and the lines copied and pasted into the active worksheet. There are, of course, other ways of saving code and making it accessible more easily, but some aspects of that process are platform dependent and will not be discussed here.

Exercise 1

Figure 1 of Section 6.6 in the text depicts a saw-tooth function called a "rasp." Plot the first, fourth, and tenth Fourier sine approximations to this function.

The rasp is generated by the periodic extension of the function f(x) = x, for x in the interval [-1,1]. To get Maple to plot the periodic extension of f(x), use the function PE defined in the Introduction above. First, be sure to define f(x) with Maple's arrow notation, thereby making f a function, not an expression.

```
>   f := x -> x;
```

Define the function whose name is "rasp" as the periodic extension (hence, PE) of the function whose name is f. Do this by applying the **PE** operator to f, being sure to terminate the command with a colon (:) since the output will look strange, and probably unintelligible. The arguments to **PE** are the function to be extended, and the domain of the function being extended.

```
>   rasp := PE(f,-1..1):
```

Obtain a graph of the rasp on the interval [-3,3], assigning the plot to a variable for use later. The plot option *discont = true* signals Maple to observe the discontinuities in the function, and tells it not to connect across the jumps.

```
>   f1:=plot(rasp(x),x=-3..3, discont=true, color=black, scaling=
    constrained,thickness=3):   f1;
```

Obtain the Fourier sine series coefficients $b_n = [\frac{2}{l}] \int_0^l f(x) \sin(\frac{n\pi x}{l}) \, dx$. Here, $l = 1$ and f(x) = x. An integral can be entered into Maple with the **Int** command which stores the integral as an unevaluated symbol. If instead, the integral is entered with the **int** command, the evaluation of the integral is immediate. Here, **Int** is used so that the integral will be displayed completely.

```
>   q := 2*Int(x*sin(n*Pi*x),x=0..1);
```

4.3 Maple On Line

To evaluate an integral that has been entered with **Int**, apply Maple's **value** command to the integral.

```
>   q1 := value(q);
```

We wish to simplify this expression. For example, $\sin(n\pi)$ is zero whenever n is an integer. We will tell Maple that n is an integer with its assume command. However, that will cause Maple to attach a tilde (~) to each n it prints thereafter. Suppressing the attachment of the tildes can be done either interactively from the Options menu (Options, Assumed Variables, No Annotation) or from the command line with the following **interface** command.

```
>   interface(showassumed=0);
```

Now, use the **assume** command to tell Maple that n is an integer.

```
>   assume(n,integer);
```

If the Fourier coefficients are now simplified, they will appear much like they would if the calculation were done "by hand." Note how, in the interest of simplicity, we assign the result to the name "b" and not to something that tries to reflect the dependence on n.

```
>   b := simplify(q1);
```

The Fourier approximations are simply partial sums of the Fourier series. The first approximation, p1, is just $b_1 \sin(\pi x)$ and the tenth one is $p10 = \sum_{n=1}^{10} b_n \sin(n \pi x)$. We can obtain these expressions in Maple by using its **sum** command.

```
>   p1:=sum(b*sin(n*Pi*x),n=1..1); p4:=sum(b*sin(n*Pi*x),n=1..4);
    p10:=sum(b*sin(n*Pi*x),n=1..10);
```

Finally, these three partial sums can be plotted with a single **plot** command by grouping the functions in a list. Colors can be assigned to the functions by the option $color = [...]$, with matching colors being listed in the order of the functions to which they are being ascribed. Assigning the plot to a variable allows merging, via the **display** command of the *plots* package, the approximations with the graph of the rasp created above

```
>   f2 := plot([p1,p4,p10],x=-3..3,color=[red, green, blue]):
    display([f1,f2]);
```

Exercise 2

Let $g(x) = \begin{cases} -1 & x < 0 \\ 1 & 0 \leq x \end{cases}$, a piecewise defined function. Obtain Fourier sine approximations with 4, 8, and 20 sine functions.

The periodic extension of g(x) is called a "square wave." Note the "ear-like" peaks which appear in the graph of the partial sums at the discontinuities of f(x). These peaks are referred to as the "Gibbs phenomenon." They are quite pronounced, even after twenty terms of the Fourier series. Their existence shows that it takes a very high fidelity amplifier to reproduce a square wave accurately. For this reason, square waves are sometimes used to test the fidelity of an amplifier.

Begin by defining g(x) as a piecewise function on the interval [-1,1]. Use Maple's **piecewise** function which permits the definition a function with multiple formulas. Define g as a function by using the arrow notation.

```
>  g := x -> piecewise(x<0,-1,x>=0,1);
```

Check the behavior of g(x) by plotting it, again using the plot option *discont = true* so that jumps in the function are not connected.

```
>  plot(g(x),x=-1..1,discont=true, color = black);
```

Define G as the periodic extension of the function g. Use the PE code detailed in the Introduction. Again, end the command with a clon (:) since the echo will probably not make much sense to the typical student of linear algebra. Plot G on the interval [-3,3], assigning the plot to a variable for use later.

As in Exercise 1, the Fourier sine coefficients b_n are computed by integrating g(x). Enter the defining integral using **int** and g(x). The presumption here is that the **interface** and **assume** commands are still operative from Exercise 1. If not, re-execute those commands.

Exercise 3

Obtain the thirtieth partial sum of the Fourier sine series for the function $f(x) = \frac{1}{1+x^2}$, $-1 \leq x \leq 1$, then graph the periodic extension of f(x) and the Fourier approximation. Explain why the graphs don't agree. (See Exercises 6 and 7 of Section 6.6).

Exercise 4

Obtain a Fourier cosine series for the function in Exercise 3. Plot the first, fourth, and eighth partial sums.

4.4 On Line

Restart Maple to clear its memory of all defined variables, and re-initialize by loading the *linalg* and *plots* packages.

Let $R_x(\frac{\pi}{6})$ be the matrix of a counterclockwise rotation, around the x-axis and through an angle of $\frac{\pi}{6}$ radians. Let $R_y(\frac{\pi}{4})$ be the matrix of a counterclockwise rotation, around the y-axis and through an angle of $\frac{\pi}{4}$ radians. Let $A = R_x(\frac{\pi}{6}) R_y(\frac{\pi}{4})$. Since the product of two orthogonal matrices is orthogonal, A is orthogonal. The purpose of these exercises is a demonstration that A defines a rotation about a fixed axis and through a particular angle. See Figure 5 in Section 4.4 of the text..

Points on this axis remain fixed under the rotation. Thus, if **X** is on the axis of rotation, it will satisfy A**X** = **X**, or equivalently, (A - I)**X** = **0**.

Exercise 1

Construct the matrix A as the product of the matrices $R_x(\frac{\pi}{6})$ and $R_y(\frac{\pi}{4})$.

Exercise 2

Find an **X** on the axis of rotation by using the equation (A - I)**X** = **0**. Thus, **X** is in the null space of A - I. Such an **X** can be found by applying Maple's **nullspace** command to A - I, which Maple lets us form via the syntax A - 1. The **nullspace** command returns a set of vectors, so **X** will have to be extracted from this set.

Exercise 3

Plot the line segment from -**X** to **X**. If we parametrize this line segment as t**X** we can plot it with the **spacecurve** command, letting t lie in the interval [-1,1]. Assign this plot to a variable so it can be used in a later exercise.

```
>   tX := evalm(t*X);

>   f1 := spacecurve(tX,t=-1..1, color=black, axes=boxed, scaling=
    constrained, labels=[x,y,z], labelfont=[TIMES,BOLD,14]):   f1;
```

Exercise 4

The plane P through the origin perpendicular to **X** is called the "plane of rotation." Since this plane contains the origin, it is a subspace of R^3. If the vector **X** is converted to a 1 x 3 matrix, Maple's **nullspace** command will produce a basis for the plane P (the null space of the matrix-form of **X**). Why? Obtain this basis, nameing its elements **N1** and **N2**. Plot line segments through **N1** and **N2** as was done in Exercise 3. Join this plot with the one from from Exercise 3 with **display3d** from the *plots* package, and assign the merged graphs to a variable for use later in Exercise 4. Be sure to use a 1-1 aspect ratio so that orthogonal vectors appear orthogonal.

The conversion of **X** to a matrix is accomplished by Maple's **convert** command, with "matrix" as the parameter. This will be a 3 x 1 matrix which the **nullspace** command will reject. Apply the **transpose** operator to produce a 1 x 3 matrix to give to the **nullspace** command.

Maple's **nullspace** command does not yield normalized vectors. After obtaining the basis of the null space, normalize the vectors with Maple's **normalize** command.

Show that the basis vectors are not necessarily orthogonal to each other. They are orthogonal to **X**. Use Maple's **dotprod** command to compute the dot products of vectors.

Exercise 5

The expectation should be that multiplication by A rotates elements within the plane orthogonal to the axis of rotation. To test this hypothesis, multiply both **N1** and **N2** by A. Then use Maple's angle command to find the angle between **N1** and A **N1**, and between **N2** and A **N2**.

A second multiplication by A should rotate A **N1** and A **N2** by the same amount. Verify this.

Exercise 6

Continue to explore, by visual means, the idea of rotations in the plane P. Create a plot of 15 successive applications of the matrix A to the basis vectors found in Exercise 2. If each application of A rotates these basis vectors through a fixed angle, and if they remain in P, their collective image should "show" the plane P.

Since working with exact expressions can lead to memory-consuming "expression swell," it is wise to convert the computations to the numeric

4.5 Maple On Line

domain. Map the **convert** operator, with the *float* option, onto the matrix A and the vectors **N1** and **N2**, coining new names for these numeric versions. For example, we might call the floating point versions of these quantities **B**, **NN1**, and **NN2**, respectively.

Multiplication of **NN1** by B, k-times, produces B^k**NN1**. We can produce the desired plots using Maple's seq command as follows. Finally, use **display3d** to merge these plots with the plot from Exercise 4.

```
>  s1 := seq(evalm(t*((B^k) &* NN1)),k=1..15):  s2 := seq(evalm
(t*((B^k) &* NN2)),k=1..15):
```

```
>  f4 := spacecurve({s1},t=0..1,color=blue):  f5 := spacecurve
({s2},t=0..1,color=green):
```

Exercise 7

Determine, in the plane P, the angle of rotation caused by multiplication by A.

Use Maple's angle command, but express the answer in degrees, as a floating point number.

4.5 On Line

Restart Maple to clear its memory of all defined variables, and then re-initialize by loading the *linalg* and *plots* packages.

Exercise 1

Imagine that you are an astronomer who is investigating the orbit of a newly discovered asteroid. You want to determine (a) what is the closest the asteroid will come to the sun and (b) what is the furthest away from the sun the asteroid will get. To solve your problem, you will make use of the following facts.

a) Asteroids have orbits which are approximately elliptical, with the sun as one focus.

b) In polar coordinates an ellipse with one focus at the origin can be described by a formula of the form

$$r = \frac{c}{1 + a\sin(\theta) + b\cos(\theta)}$$

where a, b, and c are constants.

You have also collected the data below; where r is the distance from the sun in millions of miles, and θ is the angle between the vector from the sun to the asteroid and a fixed axis through the sun. The data is, of course, subject to experimental error.

$$\begin{bmatrix} \theta & 0 & .6 & 1.8 & 1.4 & 2.1 & 3.2 & 5.4 \\ r & 329.27 & 313.8 & 319.49 & 310.91 & 327.88 & 374.91 & 367.49 \end{bmatrix}$$

Your strategy is to use the given data to find values of a, b, and c which make the formula for the ellipse to agree as closely as possible with the given data.. This will involve setting up a system of linear equations in a, b, and c and solving the normal equation. (Give the augmented matrices for both the original system and the normal equation.) You will then graph the given formula and measure the desired data from the graph.

Note: Once you have found values of a, b, and c, you will need to plot the orbit of the asteroid.

This can be done with the following command where r is the function that defines the ellipse.

>   p3 := plot([r,t,t=0..2*Pi],coords=polar):

Remark: As stated, this is an inherently nonlinear problem which we solve using linear equations. There are more accurate techniques based on multivariable calculus. These techniques are also considerably more complecated than our solution.

Exercise 2

Maple has a **leastsqrs** command from the *linalg* package that is easier (and better) for solving least squares problems than simply solving the normal equations. Use this command to solve the over-determined linear system developed in Exercise 1. An appropriate syntax is as below where **A** is the coefficient matrix for the system and **F** is the vector of constants on the right side of the equations.

>   leastsqrs(A,F);

Chapter 5

Determinants

Determinants are extremely useful in many contexts. You will, for example, use them constantly when you study eigenvalues and eigenvectors later in the text. In addition, you will see them used to write formulas for the solutions to many applied problems. In particular, determinants are used extensively in the study of differential equations and in the study of advanced calculus. Determinants are also used extensively in studying the mathematical foundations of linear algebra. Computers, however, do not generally use determinants for computations. Much faster and more efficient numerical techniques have been found. Thus, we will not provide any computer exercises for this chapter.

The reader should be aware, however, that Maple will compute determinants. The appropriate command is "det(A);". Incidentally, Maple uses the methods of the next section to compute determinants rather than the methods already described.

Chapter 6

Eigenvectors

6.1 On Line

Restart Maple to clear its memory of all variables, and re-initialize it by loading the *linalg* package.

In the **On Line** section for Section 5.1 we commented that virtually anything you might use determinants for, a computer would do otherwise. This includes finding eigenvalues. Algorithms for computing eigenvalues are very sophisticated, and will not be described in this text. However, we will point out that the techniques do not involve finding the characteristic polynomial and determining its roots. In fact, the numeric recipes for finding eigenvalues are so good that they are often used to find roots of polynomials!. The exercises in this section explore this idea.

Exercise 1

If A is the matrix $\begin{bmatrix} 0 & 1 \\ -b & -a \end{bmatrix}$, show (by hand) that the characteristic polynomial is $p(\lambda) = \lambda^2 + a\lambda + b$. Use this to construct a matrix A1 which has $p_1(\lambda) = \lambda^2 + 7\lambda + 1$ as its characteristic polynomial. Check that your matrix has the correct characteristic polynomial using Maple's **charpoly** command.

Note: The **charpoly** command requires that you name the variable to be used in the characteristic polynomial. Thus, an appropriate syntax might be

> p_1:=charpoly(A,lambda);

Also, Maple uses $p(\lambda) = \det(\lambda I - A) p(\lambda) = \det(A - \lambda I)$ as the definition of the characteristic polynomial. The relation between the two is that if A is (n x n), then the characteristic polynomial we compute in the text is $(-1)^n$

Use Maple's **eigenvals** command to compute the eigenvalues of A1 and hence the roots of $p_1(\lambda)$. Check your calculation by using the **solve** command.

Note: An appropriate syntaxt for the solve command is

```
> solve(p1 = 0,lambda);
```

Exercise 2

Compute the characteristic polynomial for the matrix $A = \begin{bmatrix} 0 & 1 & 0 \\ 0 & 0 & 1 \\ -c & -b & -a \end{bmatrix}$.

Use this result to obtain the roots of the polynomial $p_1(\lambda) = \lambda^3 + 8\lambda^2 + 17\lambda + 10$. Test the roots by substitution back into $p(\lambda)$.

Exercise 3

Let A1 be the matrix obtained in Exercise 3. Obtain the eigenvectors of A1, normalizing them so the first element in each is 1. What do you then notice about these eigenvectors? Use the pattern you articulate to give a general prescription of the eigenvectors of an (n x n) matrix of the form A. Prove your answer.

The eigenvactors for A1 may be computed using the following command.

```
> q := eigenvects(A2);
```

Note that there are three lists in q, and in each list there are three members. The first member of each list is the eigenvalue. The second member of each list is the algebraic multiplicity, the number of times that eigenvalue is a root of the characteristic equation. The third member of each list is a set of eigenvectors. Here, each such set contains a single eigenvector. Hence, these eigenvectors can be referenced as follows.

```
> v1 := q[1][3][1]; v2 := q[2][3][1]; v3 := q[3][3][1];
```

Exercise 4

Find a 4 x 4 matrix A whose characteristic polynomial is $p(\lambda) = \lambda^4 + 3\lambda^2 - 5\lambda + 7$. Obtain the roots of this polynomial by finding the eigenvalues of the matrix A. Check your result with Maple's **solve** command.

6.2 Maple On Line

The matrix whose characteristic polynomial is $p(\lambda)$ is known in mathematics as the companion matrix. Maple has the built-in command **companion** for finding the companion matrix for a polynomial. Maple generates a companion matrix that is the transpose of what you might find in some differential equations texts.

```
> p := x^4 + 3*x^2 - 5*x + 7;
> A := transpose(companion(p,x));
```

Find the eigenvalues of A.

```
> eigenvals(A);
```

Maple has expressed the roots of the characteristic polynomial with its "RootOf" notation. This is a shorthand for what could be large and complex expressions for the exact value of the roots. There are several options available at this point. First, apply the **allvalues** command to the RootOf structure. This will return the eigenvalues as exact values, containing complicated expressions involving radicals.

```
> q := eigenvals(A);
> q1 := allvalues(q);
```

These expressions are too unwieldy to work with. Convert them to floating point numbers with the **evalf** command.

```
> evalf(q1);
```

Another alternative is to include at least one floating point number in the matrix A. When **eigenvals** sees the float, it will compute the eigenvalues numerically.

```
> A1 := map(convert,A,float);
> eigenvals(A1);
```

Finally, solve the equation $p(\lambda) = 0$ numerically by using the **fsolve** command. For polynomial equations, this command accepts the parameter "complex" to indicate that all roots, both real and complex, are to be found.

```
> fsolve(p,x,complex);
```

6.2 On Line

Restart Maple to clear its memory of all variables, then reinitialize by loading the *linalg* package.

Create your own eigenvalue problem by constructing a 3 x 3 matrix A with prescribed eigenvalues and eigenvectors. The eigenvalues are the diagonal elements in a diagonal matrix D, while the eigenvectors are the columns of a 3 x 3 matrix P.

A simple way to construct the eigenvector matrix P is as a random matrix. Hence, define f, a function which generates random integers in the interval [-10,10].

```
> f := rand(-10..10):
```

Let P be a random 3 x 3 matrix with entries determined by f.

```
> P := randmatrix(3,3,entries=f);
```

Let the eigenvalues be 2, 2, and 3, in that order. Create a diagonal matrix with these elements on the diagonal, but assign the matrix to the name d, not D. The letter "D" in Maple is reserved for one form of the differentiation operator and Maple will not let you assign to it.

```
> d := diag(2,2,3);
```

The matrix $A = P D P^{(-1)}$ will have eigenvalues 2, 2, and 3 - in that order - and will have the columns of P as eigenvectors, in corresponding order.

```
> A := evalm(P &* d &* inverse(P));
```

Exercise 1

By computing A **X** for each column **X** of P, verify that each column of P is an eigenvector of A. Columns of P can be referenced by Maple's **col** command. Clearly, A **X** = λ **X** must hold for each eigenpair of eigenvector **X** and eigenvalue λ.

Exercise 2

Verify that the diagonal elements of D are the eigenvalues of A by using Maple's **eigenvals** command to determine the eigenvalues of A directly from A itself. There is, however, no canonical ordering for the results of this command, so Maple need not order the eigenvalues as 2, 2, 3.

```
> eigenvals(A);
```

6.2 Maple On Line

Exercise 3

By applying Maple's **eigenvects** command to A, again verify that the columns of P are the eigenvectors. The **eigenvects** command returns lists with three members in each list. These three members are first, the eigenvalue; second, the algebraic multiplicity of the eigenvalue - the number of times the eigenvalue was a root of the characteristic equation; and third, a set of eigenvectors associated with the eigenvalue in the list.

Extract the eigenvalues and eigenvectors by adroit use of the selector bracket notation []. (See Exercise 3 in the On Line exercises for Section 6.1.) Again, there is no canonical ordering for the lists produced, or for the eigenvectors associated with an eigenvalue of multiplicity greater than 1. Executing the **eigenvects** command on different occasions can result in a different ordering each time.

The eigenvectors computed by the **eigenvects** command may not "look like" the columns of P. The columns of P may be constant multiples of the corresponding vectors, or, in the case of multiple eigenvectors, the columns of P could simply be a different basis for the eigenspace associated with the repeated eigenvalue. Compare the third column of P with the eigenvector Maple found for the eigenvalue 3, determining any multiplicative factor needed to make the eigenvectors match exactly.

To show two bases $\{P_1, P_2\}$ and $\{v_1, v_2\}$ are equivalent, you need to show that linear combinations of one set of basis vectors yield the other basis vectors. Use the **rref** command on a matrix containing as its columns the first two columns of P and the eigenvectors Maple found for the eigenvalue 2. How will this show the equivalence of the bases?

Exercise 4

For the $n \times n$ matrix A, the characteristic polynomial $p(\lambda)$ has been defined in this text as $p(\lambda) = \det(A - \lambda I)$. Some texts use $\det(\lambda I - A)$, thereby making the two definitions differ by a factor of $(-1)^n$. Maple's built-in **charpoly** command for generating the characteristic polynomial uses the latter convention. The advantage of Maple's definition is that for an (n x n) matrix, the highest order term

In Maple, compare these two methods for obtaining the characteristic polynomial. The charpoly command takes as arguments the matrix A and a variable to be used in the output polynomial. Maple computes determinants via the **det** command, and also allows the syntax $A - \lambda$ as a short form of $A - \lambda I$. Finally, typing out the "name" of the Greek letter λ causes Maple to print that letter as a Greek letter.

> charpoly(A,lambda);

>   det(A-lambda);

Note that the two polynomials are just negatives of each other.

There is only one degree three polynomial with roots 2, 2, and 3, that has λ^3 as its highest degree term. What is this polynomial? (Hint: Write it as a product of linear factors and then expand.)

Note that this is the characteristic polynomial of A as found by Maple.

6.3 On Line

Restart Maple to clear its memory of all defined variables, and re-initialize by loading the *linalg* package.

In these exercises complex numbers will appear. Maple uses the letter "I" for $\sqrt{-1}$, so that the complex number z = 2 + 3 i is entered into Maple as z = 2 + 3*I. It is also useful to remember that if z = 2 + 3 i, then $\overline{z} = 2 - 3\,i$ is the complex conjugate of z. Thus, the complex conjugate of a real number x is that real number itself, since the imaginary part (the part with the i) is zero.

Maple's command for conjugating a number is **conjugate**. Its commands for extracting the real and imaginary parts of a complex number are **Re** and **Im**, respectively. In purely numeric contexts these commands usually need no additional boosts. In symbolic and exact contexts, these commands generally work only if an additional **evalc** (evaluate complex) is applied. Thus,

>   Re(2 + 3*I); Im(2 + 3*I);

but

>   Re(x + I*y); Im(x + I*y);

thereby requiring

>   evalc(Re(x + I*y)); evalc(Im(x + I*y));

Exercise 1

Use Maple's **eigenvects** command to obtain the eigenvalues and eigenvectors of the matrix A = $\begin{bmatrix} 1 & -3 \\ 1 & 1 \end{bmatrix}$ from Example 2 of section 5.3. Notice that the eigenvalues are complex conjugates, as are the eigenvectors as well.

Exercise 2

An $n \times n$ matrix A with complex entries is said to be Hermitian if the conjugate of the transpose equals A. Thus, A is Hermitian if $\overline{A^t} = A$. A moment's reflection reveals that the conjugate of the transpose equals the transpose of the conjugate, that is, $\overline{A^t} = \overline{A}^t$. Some texts denote the conjugate transpose of A by the symbol A^* so that $A^* = \overline{A}^t = \overline{A^t}$, and A is Hermitian provided $A = A^*$.

Give an example of a 3 x 3 Hermitian matrix containing as few real numbers as possible, and having no entries zero.

To verify that your matrix is Hermitian, you need to take both the transpose and the complex conjugate. Maple has the built-in **htranspose** command for this. Apply it to your matrix A, and then separately apply the **transpose** and **conjugate** commands. In Maple, the simplest way to do this is to map **conjugate** onto the transpose of A.

```
>   htranspose(A);

>   map(conjugate,transpose(A));
```

One remarkable property of Hermitian matrices is that their eigenvalues are real. Apply Maple's **eigenvals** command to your matrix A in an effort to verify the truth of this claim. You will most likely obtain large, complicated expressions for the symbolically exact eigenvalues. They might even contain the symbol "I", making it look like the eigenvalues are complex! To determine if these eigenvalues are real, you want to show that for each, the imaginary part is zero.

After obtaining the exact eigenvalues, convert them to floating point numbers by using the **evalf** command. Since **eigenvals** returns a sequence of eigenvalues, you may need to convert this return to a list (use [...]) before the **evalf** command works. The conversion to decimals of any radicals in your eigenvalues may yield expressions with very small imaginary parts. You can instruct Maple to truncate these small numbers to zero by using the **fnormal** command.

Instruct Maple to extract and simplify the imaginary part of each eigenvalue. These imaginary parts should reduce exactly to zero. Remember to use both **evalc** and **Im**, as well as **simplify**, on each exact eigenvalue.

Obtain the exact real part of each eigenvalue. These expressions will be real, but complicated. The point of the activity is for you to realize that exact values for the roots of cubic equation are unpleasant expressions to work with. Just because Maple is able to provide the roots exactly does not mean that these expressions are always useful or simple.

Finally, apply evalf directly to the exact real parts of the eigenvalues.

Compare the values to what you got when floating the unsimplified "complex" version of the eigenvalue.

Exercise 3

Change one of the entries of A from Exercise 2, making A non-Hermitian. Then, recalculate the eigenvalues. Are they again real?

Make the change to A in Maple by creating a matrix B via substitution of a new value into the matrix A. Reference an element of A, say the 1,1-element, by A[1,1], and make a substitution of a new value for such an element into op(A), not just into A. Choosing the new value to be a floating point number will mean that **eigenvals** will return floating point numbers directly. Since Exercise 2 made a thorough study of the complexity of exact calculations, work numerically in this exercise.

Exercise 4

For the matrix A of Exercise 2, compare the action of transpose and htranspose. Is there any difference if these operators are applied to matrices with just real entries?

Exercise 5

In Maple, let A be a real, symmetric 3 x 3 matrix with as many of its entries as possible distinct. Obtain $B = I + i\,A$, where I is the 3 x 3 identity matrix and $i = \sqrt{-1}$. Let $C = I + \frac{B^{(-1)}}{2}$ and compute both $C\,C^*$ and $C^*\,C$. Can you prove that what you observe is always true? (Hint: Begin with the equality $B + B^* = 2\,I$ and multiply by $B^{(-1)}$ on the left and by $[B^*]^{(-1)}$ on the right.)

Notes: To generate a random symmetric matrix in Maple, add the parameter *symmetric* to the **randmatrix** command.

```
>    A := randmatrix(3,3,symmetric);
```

The matrix B can be obtained in Maple if due note is taken of Maple's usage of I for the imaginary unit $\sqrt{-1}$, and due care is taken to distinguish between the written symbols I and i, standing respectively for the identity matrix and the imaginary unit $\sqrt{-1}$.

Since an identity matrix is a diagonal matrix with just 1's on the diagonal, the **diag** command can be used to create an identity.

```
>    B := evalm(diag(1$3)+I*A);
```

6.5 Maple On Line

The matrix C likewise requires use of an identity matrix.

```
> C := evalm(diag(1$3)+inverse(B)/2);
```

6.5 On Line

Restart Maple to clear its memory of all variables, then re-initialize by loading the *linalg*, *plots*, and *plottools* packages.

```
> restart;
> with(linalg): with(plots): with(plottools):
```

The purpose of this exercise set is to explore the relationship between eigenvalues, eigenvectors and the geometry of quadratic forms.

Exercise 1

Use Maple's **implicitplot** command from the *plots* package to obtain a graph of the ellipse defined by the quadratic equation $x^2 + 4y^2 = 1$. Be sure to use a 1-1 aspect ratio so that there is no distortion in the scaling. Then, obtain A, the matrix of the quadratic form defined by this same equation. This can be done by typing in A, by clever use of partial differentiation, or by Maple's **hessian** command. Obtain the eigenvalues and eigenvectors of A.

```
> q := x^2 + 4*y^2 = 1;
> implicitplot(q,x=-1..1,y=-1..1,scaling=constrained);
```

By inspection, we can write $A = \begin{bmatrix} 1 & 0 \\ 0 & 4 \end{bmatrix}$. Alternatively, we can note that $A = [\frac{1}{2}] \begin{bmatrix} f_{xx} & f_{xy} \\ f_{yx} & f_{yy} \end{bmatrix}$, where f(x,y) is the left hand side of the defining quadratic equation, and subscripts denote partial derivatives. Thus,

```
> f := lhs(q);
> fxx := diff(f,x,x); fxy := diff(f,x,y); fyx := diff(f,y,x);
  fyy := diff(f,y,y);
```

The array of second partial derivatives of the form $\begin{bmatrix} f_{xx} & f_{xy} \\ f_{yx} & f_{yy} \end{bmatrix}$ is called the hessian matrix, and is returned in Maple by the **hessian** command.

```
> hessian(g(x,y),[x,y]);
```

Hence, the matrix of a quadratic form is simply one-half the hessian matrix. The simplest way to get the 1/2 into the hessian is by including it in the quadratic function. Else, an **evalm** is needed to multiply the hessian matrix by that 1/2.

>   A := hessian(lhs(q)/2,[x,y]);

The eigenvalues and eigenvectors can be obtained by use of the **eigenvects** command.

We next seek to relate the eigenvalues and eigenvectors to the lengths of the semi-major and semi-minor axes. From the graph, these axes are 1 and 1/2 respectively. The eigenvalues are 1 and 4, respectively for eigenvectors that have the directions of the ellipse's axes. Hence, the scale factors by which to multiply the eigenvectors to get the semi-major and semi-minor axes are $\frac{1}{\sqrt{\lambda_k}}$, where k = 1, 2.

Exercise 2

Using Maple's **implicitplot** command, obtain a graph of the function defined implicitly by the quadratic equation $2y^2 + xy + x^2 = 1$. Be sure to use an aspect ratio scaled to 1-1. It may also be edifying to increase the number of points used in the plot by adding the parameter *numpoints* = 1000 to the **implicitplot** command. Assign the plot to a variable so that it can be re-used in subsequent exercises.

Exercise 3

For the quadratic equation in Exercise 2, form the matrix A of the quadratic form defined by the equation. Use Maple's **hessian** command, and check the result by taking partial derivatives of the left hand side of the defining quadratic function. Obtain the eigenvalues and eigenvectors of A via Maple's **eigenvects** command. Extract, and give unique names to, the eigenvalues and eigenvectors. Obtain floating point approximations for the eigenvalues, and normalize the eigenvectors to have length 1 with Maple's **normalize** command. Apply Maple's **radsimp** command to the 2-norm (computed via **norm**) of each normalized eigenvector to verify that each indeed has length 1. Use the **arrow** command from the *plottools* package to superimpose the normalized eigenvectors on the graph of the ellipse. Define each arrow separately. The arrow command takes five parameters: a list of coordinates for the tail, here the origin; a list of coordinates for the point, here the entries in the normalized eigenvectors; then three sizing parameters which experiment shows are well chosen as .05, .1, and .05. Color each

6.5 Maple On Line

arrow differently. Then use the **display** command from the *plots* package to merge the graph of the ellipse and the two arrows into one graph. To convert a vector to a list, use the **convert** command with option *list*. Thus, if the eigenvectors are named V1 and V2, you might eneter

```
> a1:=arrow([0,0],convert(V1,list),.05,.1,.05, color=green):
> a2:=arrow([0,0],convert(V2,list),.05,.1,.05, color=blue):
> display([p1,a1,a2],scaling=constrained);
```

Print a copy of this final graph of the ellipse with the normalized eigenvectors.

Exercise 4

On the plot printed in Exercise 3, draw in the axes determined by the eigenvectors. On these axes put tick marks every quarter unit, noting that each eigenvector is one unit long. Use a ruler to guarantee the accuracy of your tick marks. According to the general theory, the ellipse should cross the new axes at points whose coordinates in the system determined by the eigenvectors are

$(\frac{1}{\sqrt{\lambda_1}},0)$, $(-\frac{1}{\sqrt{\lambda_1}},0)$, $(\frac{1}{\sqrt{\lambda_2}},0)$, $(-\frac{1}{\sqrt{\lambda_2}},0)$, where λ_1 and λ_2 are the eigenvalues of A. Verify this by estimating the appropriate coordinates from your graph.

Exercise 5

Verify in general that the scale factors $\frac{1}{\sqrt{\lambda_k}}$, k = 1, 2, indeed convert normalized eigenvectors into vectors of precisely the length of the semi-major and semi-minor axes. Begin by obtaining floating point values of $\frac{1}{\sqrt{\lambda_k}}$, k = 1, 2. Then scale the normalized eigenvectors by these factors. Next; use Maple's arrow command to build a plot of the ellipse, and the newly scaled eigenvectors. A plot of the ellipse, and these new basis vectors should show that with this scaling the vectors coincide precisely with the semi-major and semi-minor axes.

Exercise 6

Find the equation of an ellipse, centered at the origin, and for which the major axis is 4 units long and lies along the line determined by the vector $v_1 = \begin{bmatrix} 3 \\ 4 \end{bmatrix}$, and for which the minor axis is 2 units long. (Hint: If you can figure out the eigenvalues and eigenvectors, you then can find matrices Q and D for which $A = Q D Q^t$, where A is the (symmetric) matrix for the quadratic form corresponding to the ellipse.) Graph the ellipse and the scaled eigenvectors for A, demonstrating that the proper scaling of the eigenvectors gives them the lengths of the semi-major and semi-minor axes.

Since the axes of the ellipse are orthogonal, you need to get a vector orthogonal to the vector v_1. This can be done in the plane by interchanging the x- and y-coordinates, and negating one component of the resulting vector. Even a casual inspection reveals why this works.